Dr Karl Kruszelnicki

50 shades of
Grey Matter

MACMILLAN

First published 2012 in Macmillan by Pan Macmillan Australia Pty Limited

1 Market Street, Sydney

National Library of Australia

Cataloguing-in-Publication data:

Kruszelnicki, Karl, 1948-
50 shades of grey matter / Karl Kruszelnicki.
9781742611389 (hbk.)

Science—Popular works.
Science—Miscellanea.

500

Cover, internal design and typeset by XOU Creative, www.xou.com.au.

Cover photo of Dr Karl by David Stefanoff.

Internal illustrations by Douglas Holgate.

Printed by McPherson's Printing Group.

Papers used by Pan Macmillan Australia Pty Ltd are natural, recyclable products made from wood grown in sustainable forests. The manufacturing processes conform to the environmental regulations of the country of origin.

This book is dedicated to the Higgs Boson
– the discovery of your existence is MASSive

TABLE OF CONTENTS

MARSHMALLOWS, MONEY AND MUNCHIES

What's the link between marshmallows, money and the munchies? Willpower and self-control!

Back in 1968, Walter Mischel from Stanford University in California came up with the now-famous "Marshmallow Test". It tested self-control in some 600 children aged between four and six.

MARSHMALLOW TEST ON KIDS . . .

The kids were each individually offered a treat: a marshmallow, a pretzel, or an Oreo cookie. If they wanted, they could eat the treat immediately. But if they waited 15 minutes and didn't eat the single treat, they would then be rewarded with an additional treat. As you might expect, the older kids were better at waiting longer.

Looking at the whole group, a small minority didn't even think about waiting and immediately ate their treat.

> In one dramatically effective self-distraction technique, after obviously experiencing much agitation, a little girl rested her head, sat limply, relaxed herself, and proceeded to fall sound asleep.

Most of the kids tried to hold off – and most failed. About one third were able to wait it out successfully for a full 15 minutes and be rewarded with their additional treat.

It was very hard for the kids. Some covered their eyes or turned around so that they couldn't see the tempting treat, others kicked a desk or stamped their feet on the floor or even pulled their hair to divert themselves from thinking about it.

Others stared into the mirror, or began talking to themselves.

The more they avoided looking at the treat, the more successful they were at waiting the full 15 minutes. Rather than stoically "willing" themselves to stare down temptation, they simply engaged in other activities to avoid looking at the treat. Some sang songs, while others hid their heads in their arms, or prayed to the ceiling. Dr Mischel noted, "In one dramatically effective self-distraction technique, after obviously experiencing much agitation, a little girl rested her head, sat limply, relaxed herself, and proceeded to fall sound asleep."

Mind you, once they had managed to successfully distract themselves for the full 15 minutes, they didn't wait any longer and immediately rewarded themselves by eating both treats.

. . . YEARS LATER

Originally, Dr Mischel had no intention of doing any follow-up. But his three daughters went to the same school as some of the four-to-six-year-old kids in his study. As part of idle dinner conversation, he would ask his daughters about these kids. As the years rolled by, he thought he could see a link between their ability to wait for the second marshmallow and how well they did in school.

So he then followed up the kids as they grew up (1981, 1990 and 2011). The results were amazing.

In the USA, the SAT exams are standardised tests for admission to university, with a maximum possible score of 2400. The students who as kids could wait 15 minutes for their treat had an SAT score that was 210 points higher than those of the children who could wait only 30 seconds. These kids also ended up being more successful and popular at school and at work, and more respected by their co-workers.

> The students who as kids could wait 15 minutes for their treat had an SAT score that was 210 points higher than those of the children who could wait only 30 seconds.

The ones who couldn't wait 15 minutes were more likely to have behavioural problems, both at home and in school. They were less likely to have done well at school, less able to deal with stress, and more likely to get fatter and to have more personal problems. They were also more likely to have been arrested, and to have problems with drugs.

Dr Angela Duckworth from the University of Pennsylvania found a similar result with eighth-grade students. She gave them a choice between a dollar right then, or two dollars one week later. The ones who could wait also did better at school.

Surprisingly, this ability to delay gratification was a much better predictor of the students' academic performance than their IQ (about two times better).

It's Not Just Marshmallows

If you can hold out for 15 minutes so that you get two marshmallows instead of one, then in later life you will probably do your homework before watching TV.

Dr Mischel found that children from poor families tended to be worse at delaying gratification. He says, "When you grow up poor, you might not practise delay as much. And if you don't practise then you'll never figure out how to distract yourself. You won't develop the best delay strategies, and those strategies won't become second nature."

But Dr Mischel found a workaround. For example, he taught the kids to pretend that the tempting treat was not really a treat, but only a photograph surrounded by an invisible frame. This simple trick dramatically improved their self-control.

SELF-CONTROL

Part of what makes up self-control is your ability to change what you do and think. So it covers how you guide your thoughts and emotions, and how well you perform your duties and tasks.

We used to think of self-control or willpower as being some kind of "moral attribute", with "stronger" people having more while others had less. But now we think of it as being more like a muscle. It can be overworked and tire out, it can get stronger with exercise, and it can be recharged after a rest and a feed.

First, even though you don't realise it, you use your self-control throughout your day. Dr Wilhelm Hoffman, currently at the University of Chicago, monitored 200 German adults by getting them to wear a beeper. When the beeper went off, they would report what they were doing at that exact moment. It turned out that his subjects were spending an amazing three to four hours every day simply resisting desires and temptations, by using their willpower and self-control.

Second, willpower is finite, and exists in a single "pool".

Each day, there are many different desires and temptations that assault you – rational thoughts, irrational emotions, desires to eat or to exercise and so on. But you do not have a separate "pool" of willpower for each of these. Instead, it appears as though you have one single "pool" of willpower that has to cover all the different temptations that confront you.

This "pool" of willpower is not very big. If you use up a lot of willpower in resisting temptation on one task, the amount that you have left to deal with a subsequent, completely unrelated, task is much lower.

Think of willpower as being like a muscle. In a race, athletes might conserve their strength and not use up all of their muscle power early, so that they have some left for a big finish. The same logic applies to willpower.

Third, this depletion of willpower can be halted, or even reversed, by glucose.

Studies have been done in which the subjects had to perform a series of willpower/self-control tasks. As expected, they gradually got worse as they moved from one set of tasks to the next. But if they drank regular lemonade they performed better than people who drank diet lemonade.

Satan – Get Thee Behind Me

In the Bible, the Devil comes to tempt Jesus during his 40-day fast in the desert. Eventually, Jesus says the now-famous sentence, "Get Thee Behind Me, Satan!"

Jesus chose an excellent strategy. If you deliberately remove the things that are tempting you from your immediate environment, then the strength of the temptation is weakened. You do not have an infinite amount of self-control. So if you remove or lessen the temptation, you will be able to spread your limited amount of self-control over a longer period of time.

This strategy is not Rocket Science (but some people might consider it Divine Intervention). It can be learned at an early age, and then be used to lessen temptation. Unfortunately, some people are never taught this.

LIVE LONG AND PROSPER

So what about people who need to exert continual self-control so that they do not overeat? On one hand, they need some glucose to top up their "pool" of willpower. But eating to get their glucose adds to the risk of actually overeating. It's a vicious circle. That's part of the reason why diets are devilishly difficult – and why 85 per cent of all diets fail.

The way around this is for dieters to have a lifestyle change. By changing the way they view food and meals, so that they select food that is both healthy and satisfying, they get a fuller tummy, which boosts their willpower and helps protect them from overeating. This is a real win-win situation.

Getting back to Dr Mischel, he has some simple advice on teaching self-control. "We should give marshmallows to every kindergartner. We should say, 'You see this marshmallow? You don't have to eat it. You can wait. Here's how.'"

Sirens – Get Thee Behind Me

We all have different shortcomings, or weaknesses, in our personality. We can't always fix them, but we can work around them or outsmart them.

If you love a certain fatty food, deliberately try *not* to buy it in the shop or supermarket. If you succeed, it won't be in your pantry at home, tempting you. (You should think of this love for fatty foods as an excellent survival tactic, not a "weakness". It's a direct result of 200,000 years of evolution, from a time when food was scarce.)

In Greek Mythology, Odysseus (or Ulysses, as the Romans called him) knew that he could not resist the song of the Sirens. He cleverly had his crew tie him to the mast so that he couldn't obey their haunting song and be lured to his death on the rocks. He also had his crew block their ears so that they could hear neither the Sirens' song nor his entreaties to release him – and so avoided the Sirens (and Doom).

2

WHY IS THE SKY BLUE?

"Why is the sky blue?" is one of the first questions that a small child asks their parents. The perpetual beauty of the blue sky wraps the entire sunlit Earth. But where does this bottomless, deep blue colour come from?

Some people wrongly believe that the blue colour of the sky is just a reflection of the blue colour of the ocean, which covers two-thirds of the planet. But the correct answer is that the sky gets its blue colour from white sunlight being scattered by the tiny molecules that make up our atmosphere.

WHAT IS SCATTERING?

The Really Deep explanation of why the sky is blue is tricky. Albert Einstein wrote, in 1911, that Relativity had to be used to fully understand why the sky is blue.

So it's OK to give the easier explanation.

It begins with understanding a phenomenon called "scattering" of light. Scattering means that the light hits some kind of particle, is absorbed, and then is emitted. However, between being absorbed and emitted, the light is usually changed. This change could be the brightness of the light, or the angle at which it is emitted, or even its colour.

> Albert Einstein wrote that Relativity had to be used to fully understand why the sky is blue.

What happens to the emitted light depends on the size of the particle that absorbs and then emits it. Gustav Mie was the physicist who gave us the first deep understanding of scattering, back in 1908. He dealt with light being scattered by particles of all sizes, from the very large (such as dust particles) to the very small (such as individual molecules of gas).

But before Gustav Mie's work, it was John William Strutt, third Baron Rayleigh, also known as Lord Rayleigh, who solved the "Why Is the Sky Blue?" problem. He worked out what happens when light is scattered only by very small particles, such as molecules of oxygen or nitrogen.

LIGHT BUMPS INTO GAS MOLECULES

Lord Rayleigh was an intellectual heavyweight, winning the Nobel Prize for Physics in 1904. But it was earlier than that, back in the 1870s, that he gave us the now-famous Rayleigh Scattering Effect, which explains why the sky is blue. Rayleigh Scattering works when the size of the particle is less than about 3 per cent of the wavelength of light. Molecules of nitrogen and oxygen are a few thousand times

> For the sunlight entering our atmosphere, it's as though it travels in a vacuum, missing the vast majority of particles, until it hits one (usually oxygen or nitrogen) and gets "scattered".

smaller than the wavelength of light, so the Rayleigh Scattering Effect is definitely happening when sunlight enters the Earth's atmosphere.

The atmosphere is made from various gases – about 77 per cent nitrogen, 21 per cent oxygen and 1 per cent argon, with several other gases (including carbon dioxide) making up the remaining 1 per cent.

These particles are very small and, relative to their size, separated from each other by great distances. For the sunlight entering our atmosphere, it's as though it travels in a vacuum, missing the vast majority of particles, until it hits one (usually oxygen or nitrogen) and gets "scattered". This scattered light then continues travelling in a vacuum until it hits another particle in our atmosphere. Each time the light hits a particle there's a time delay, during which the light is absorbed and then emitted.

RAYLEIGH SCATTERING

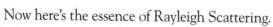

Now here's the essence of Rayleigh Scattering.

First, your eyeball will register seeing a colour only if some light of that colour lands on your retina, inside your eye.

Second, the emitted light does not pass through the molecule still travelling in its original direction – as if nothing had happened. Instead, mostly it comes out at right angles to its original direction of travel.

Third, the "right-angle bending effect" is very much stronger for blue light than for red light. So most of the light emitted from a molecule of oxygen or nitrogen is predominantly blue. There is some red light emitted, but it mostly goes straight ahead. This means that you don't see much red light at all.

RED AHEAD, BLUE BENT

So white sunlight will hit a gas particle, but most of what comes out is blueish light – and at right angles to the original direction. This blueish light will then travel in a straight line until it hits another particle, and the whole absorb–emit–right-angle–blue thing happens again. So the light that came from the sun will hit a gas particle high in the atmosphere, and then will stutter in a zigzag fashion down towards the ground. It's kind of like a ball rolling down a staircase. Each time the light hits a particle, the light comes out mostly at right angles to its previous direction of travel – and each time, it's more blue in colour. Eventually, some light that has been through thousands of these absorb–emit interactions with thousands of particles will land in your eyeball – and then, and only then, do you get the impression of seeing blue.

So the sky is blue because of Rayleigh Scattering by the nitrogen and oxygen molecules in the atmosphere.

In the movie *The Wizard of Oz* Dorothy sings about skies being blue over the rainbow. It's a nice song, but blue skies happen in most places, not just over the rainbow . . .

No Blue Sky in Paintings = Ultramarine

Before 1300, the sky was simply not painted as blue. This was mostly because the only blue pigment until then was ground-up lapis lazuli. "Lapis" is Latin for "stone", while "lazuli" comes from Latin and Arabic words meaning "heaven" or "sky". It's an intensely blue, semi-precious stone – and was mined in Afghanistan some 5000 years ago.

It was incredibly expensive, because it came from "beyond the seas", in the sense of "far away" or "over the horizon". And that's why the word "ultramarine" ("ultra" meaning "beyond" and "marine" meaning "seas") was given to this intense shade of blue.

3

MIRACLE
FRUIT

Finally, after centuries of waiting, a scientific mystery has been solved. We now understand how Miracle Fruit can change the taste of sour into sweet.

Miracle Fruit – what is it?

THE MIRACLE OF MIRACLE FRUIT

> I swallowed the pulp and after a minute the chef gave me a lemon to chew. Amazingly, it tasted like candied lemon!

My first introduction to Miracle Fruit (sometimes called the Miracle Berry) was in a radio station. The guest after me was a chef and in the studio, live on air, he gave me a small red berry to pop into my mouth. I bit into it and spat out the small black seed. As I chewed the Miracle Fruit, the mashed-up pulp mixed with my saliva and then spread all across the inside of my mouth and tongue. It tasted fairly neutral – certainly not sweet.

I swallowed the pulp and after a minute the chef gave me a lemon to chew. Amazingly, it tasted like candied lemon! Bitter vinegar tasted like sweet apple juice! Anything acid tasted sweet – and not just sweet but very, very sweet. Strangely, banana still tasted just like banana.

So here's a clue – the Sour-to-Sweet Effect happens only in the presence of something rather acidic.

MIRACLE FRUIT 101

Miracle fruit comes from a shrub from tropical West Africa, *Synsepalum dulcificum*. It grows up to 6 metres tall, and has brown flowers with 2–3 centimetre red berries – yup, they're the Miracle Fruit. The locals have long used it to add sweetness to palm wine or beer that's gone sour, or to make stale, acidified maize bread more palatable. The effect lasts for a few hours.

Miracle Fruit first came to the attention of Europeans in 1725, when the explorer Chevalier des Marchais came across it on a botanical expedition in tropical West Africa. In 1852, the chemist

W. F. Daniell was the first European to scientifically study it, and he gave it the name "miraculous berry".

It took until 1968 for two separate groups of scientists to isolate the berry's active ingredient. It turned out to be a chemical that was about 86 per cent protein (with about 191 amino acids), and about 14 per cent carbohydrate (sugars such as mannose, galactose and fucose). This active ingredient was given the name "miraculin".

> The locals have long used it to add sweetness to palm wine or beer that's gone sour, or to make stale, acidified maize bread more palatable.

Alternative Sweetener?

Soon after miraculin was isolated, Robert Harvey, an American biomedical postgraduate student, became aware of its wonderful sweetness property. At the time, the available artificial sweeteners (which have sweetness, and virtually zero kilojoules) had a slightly noticeable aftertaste. But miraculin did not.

Harvey tried mightily to market it as an alternative sweetener – and one that was based entirely upon a natural product. But in 1974, just as he was about to launch it, the US Food and Drug Administration refused to classify it as "generally recognized as safe", despite West Africans having eaten Miracle Fruit for centuries with no problems. Robert Harvey could not afford the several years of testing needed, so miraculin didn't make it into the marketplace.

HOW TASTE WORKS

How does Miracle Fruit turn sour into sweet?

When you "taste" something, you mostly use your tongue, which sends information to your brain. Your tongue has a bumpy surface, because it's covered with 2000–8000 raised taste buds of different sizes and shapes. Some of them look like tiny mushrooms, others like mini-volcanoes, and some like cylindrical shrubs.

When you eat something, the chemicals in the food land on your tongue – and its several thousand taste buds. Each taste bud holds about 50–100 taste receptor cells. Each of these taste receptor cells will respond to only one of the five basic tastes – sweet, salt, sour, bitter and umami. So, for example, when a sweet-tasting chemical lands on a taste receptor cell that responds specifically to sweetness, then that receptor cell gets activated and sends a signal to the brain.

Flavour-tripping Party

Some people run "Flavour-tripping Parties". You pay your money ($10–$15 or so), turn up, eat your single red Miracle Fruit, and then, after a minute or so, start browsing the unusual range of foods that have been laid out.

So Guinness, when you add some acidic lemon sorbet, suddenly tastes like a chocolate milkshake. Tabasco sauce now tastes like hot doughnut glaze. Lemons taste like candied lemons. Goat cheese now tastes like cheesecake – and so on.

But, to understand the secret of Miracle Fruit, we need to go into a bit more detail.

Most chemicals "talk" or interact with your cells via their shape. It's a lock-and-key kind of thing. In your front door, if the key is the exact correct shape for your lock, then the lock opens and you can get into your house.

In the same way, all cells in your body, including the hundreds of thousands of taste receptor cells on your tongue, are covered with microscopic receptors – they're like tiny locks. These receptors are hollow. If a chemical can fit into that receptor and make a good fit, then the cell will respond. In the case of taste receptor cells, they respond by telling your brain that you have just tasted one of the five basic tastes.

Umami

You've probably heard of the Four Basic Tastes – Sweet, Sour, Salt and Bitter. Well, let me introduce you to the Fifth Basic Taste – Umami.

Umami is a Japanese word meaning "pleasant savoury taste". It was first accepted as the Fifth Taste in 1985, by the scientists at the first Umani International Symposium in Hawaii, as the taste of "glutamates" and "nucleotides". It's poorly described as being "meaty" or "brothy", leaving a mild but lasting aftertaste.

It's found in human breastmilk, fish, shellfish, mushrooms, spinach, green tea, and fermented products such as soy sauce.

HOW MIRACLE FRUIT WORKS

Now here's the trick of Miracle Fruit. It's a two-step process.

Remember that the active ingredient in Miracle Fruit is a chemical called miraculin. It's a mid-size chemical. It has a molecular weight of about 25,000 Daltons – about 5 times heavier than insulin, or 1,400 times heavier than water.

When you eat Miracle Fruit, the first step is that the miraculin slots loosely into the sweet receptors on your tongue. Because if it doesn't fit exactly, it doesn't work. So the miraculin does not send the impression of sweetness to your brain, even though it has slotted into the sweet receptor. In fact, if you eat something sweet, your brain won't know. This is because the miraculin is already sitting inside all of the sweet receptors, and so it blocks the sweet food you have just eaten from slotting into those receptors.

The second step is when you eat or drink something acid.

The miraculin suddenly changes shape so that it is now a good fit inside the receptor. In fact, it's better than just a good fit – it's an almost perfect fit.

How perfect? Miraculin binds to the sweet receptor a million times better than the artificial sweetener aspartame, and 100 million times better than sugar. So your brain is being told that there's an incredible amount of sweetness on your tongue, even though you are eating a very sour lemon.

The miraculin will hang in there, sitting on the sweet receptors on your tongue for over an hour, before it gets washed off by your saliva.

Scientists are using genetic engineering to modify other plants (such as fruits or vegetables) to grow miraculin. This natural product could be used (say) for people with diabetes, to satisfy their sweet tooth without the extra calories or kilojoules that sweetness normally brings, or to make food taste nice again for people on chemotherapy.

PARTY IN MOUTH?

But, as the great physicist Richard Feynman said, "You can't fool nature".

Sure, you've had a wonderful party going off in your mouth as you gobbled down lemons, vinegar, pickles and beer, glorying in the strange new taste sensations from your easily fooled brain. But your stomach is not fooled. It's full of lemons, vinegar and all kinds of other food items that your taste buds would normally warn you should not be mixed.

Get ready for the Mother of All Tummy Aches, thanks to the Sweet Lie of Miraculin . . .

4

DOORWAYS
AND FORGETTING

Your memory is a funny thing. Have you ever walked out your front door and suddenly realised you can't remember where you parked your bicycle or car? Or have you ever gone to the kitchen because you were thirsty – and then forgotten why you went there? And when you do eventually wander back with a glass of water in your hand, do you forget what you were working on before you got thirsty?

What's going on? Are you coming down with Early-Onset Dementia? Nope, the real problem is that pesky doorway – it's the reason why you can't remember.

THERE'S TOO MUCH GOING ON . . .

Now, for a bit of essential background, you have to remember that the universe around you is Really, Really Big. There's so much information around you that your brain has to block some of it out. (Don't walk through a doorway while you read this . . .)

Another factor is that the universe is a dangerous place. You've got to be continually on the lookout for threats – especially when you go into a new location.

Those naughty doorways make you forget the past, because they alert you to the fact that you have entered a new environment.

> The universe is a dangerous place. You've got to be continually on the lookout for threats – especially when you go into a new location.

At least, that's the result of a psychology experiment by Dr Radvansky and colleagues from the University of Notre Dame in Indiana in the USA. He was exploring these memory lapses in terms of what psychologists call an "Event Horizon Model". One component (or aspect) of this Model is a belief that events are separate, and that we humans process them one at a time. Another aspect is that if you're concentrating on something (say, a new location), it will take up most of your attention, and everything else gets pushed into the background. And finally, this Model says that different parts of your memory system (for example, new location, old location) compete with each other to try to retrieve information.

Science and Psychology

The famous physicist Ernest Rutherford had a rather biased view of Physics, and said words to the effect of: "Physics is the only real Science, everything else is stamp collecting."

However, when Rutherford was awarded the 1908 Nobel Prize in Chemistry, he said, "I must confess it was very unexpected, and I am very startled at my metamorphosis into a chemist."

Taking a wider view of Life, the French poet Paul Valéry said, "The purpose of psychology is to give us a completely different idea of the things we know best." (On the other hand, Paul Valéry also said, "The future is not what it used to be.")

THE EXPERIMENTS

In Dr Radvansky's study, each object that the volunteers had to remember came in a combination of different colours and shapes. There were ten different colours, and six different shapes, so you could have a blue cube, a red wedge, a yellow cone and so on. That's a total of 60 different items.

The first experiment was a virtual one on a smallish computer screen. In the computer game, the volunteers would pick up any one of these 60 items from a table and put it on to another table. Once they picked it up, the item would vanish into a virtual backpack so they couldn't see it any more. Sometimes the table that they placed

The famous physicist Ernest Rutherford had a rather biased view of Physics, and said words to the effect of: "Physics is the only real Science, everything else is stamp collecting."

the item on was in the same room, but sometimes it was in a different room. If they had to go through a *doorway* into a *different* room, they were more likely to forget what object they were carrying.

The second experiment was pretty well the same, except that everything was real – the tables, the items and the rooms. The result was the same – walking through a *doorway* into a *different* room gave them more memory lapses.

But what was the cause – the doorway, or the fact that the room they ended up in was different?

So the third experiment had the volunteers going through a few doorways, and then back into the same room that they started in. Going back into the same room did not improve their memory.

So it seems that the memory lapse was caused by the act of passing through the doorway.

Complicated Memory

Memory is so much more complicated than just the Event Horizon Model.

Clearly, another very important factor in memory is the context, or location or situation, in which you know (or remember) something. This helps explain why you can't recognise your sister-in-law if you unexpectedly meet her at the airport rather than at your brother's house.

WHY DO DOORWAYS MAKE YOU FORGET?

Why does a doorway wipe your memory? Why does going out the Front Door make you forget if you have locked the Back Door?

Why on Earth do we have a memory system that forgets stuff as soon as we change our environment?

Actually, forgetting the old might have a survival advantage. A new environment is unknown, and might be dangerous. So it makes sense to forget the old environment and put all your attention on the new. Forget about the killer gorillas in the forest you just left – start looking for the killer lions on the open plains you have just entered. What can you do, apart from banning all doorways and living only on the open fields, or in open-plan houses?

Maybe you need to ensure that the Doorways of Perception are aligned to the Windows of your Mind . . .

5

BREASTMILK

The year 2012 is the 50th anniversary of the very first silicone breast implant. It was performed in Houston, Texas, and Timmie Jean Lindsey still has the implants. Even though they have ruptured, she doesn't want them removed.

While breasts do look pretty good, looks aren't everything. Their main function is to feed a newborn child. Even today breastmilk is still surprising us. The molecular biologists, biochemists and geneticists who attended the meeting of the International Society for Research in Human Milk and Lactation in Lima, Peru, had plenty of bombshells.

For example, I always thought that breastmilk was served up to the baby sterile – with no bacteria in it. But it turns out that up to 600 different species of bacteria can be cultured from one single sample of breastmilk. It seems that breastmilk is similar to a complex cultured yoghurt, in the sense that many different species of live bacteria contained in it do stuff we still don't understand.

Another major surprise are the nutrients in breastmilk. Basically, breastmilk contains fats (about 4 per cent), proteins, carbohydrates and a whole bunch of other stuff. The other stuff includes vitamins A, C, E and K, as well as essential minerals. Besides the well-known lactose (a carbohydrate), there are also some mysterious oligosaccharides. (Oligosaccharides are also carbohydrates that are made of short chains of various sugars joined together.) On a percentage basis, there is roughly as much oligosaccharide as there is protein in breastmilk, and more than there is fat.

> **Up to 600 different species of bacteria can be cultured from one single sample of breastmilk.**

Now here is a very weird thing: the baby cannot digest these oligosaccharides, so why are they there? Not all mothers make the same set of oligosaccharides, but we do know that they tend to vary roughly following the mother's blood group.

Perhaps they are there to feed the rapidly changing populations of bacteria in the baby's gut. At this stage, we simply don't know.

Human Breasts

As you would expect, all mammals have mammary glands. In humans, boys and girls have identical mammary glands until puberty, when hormones make them change shape.

Humans are the only mammal where the females develop breasts at puberty and then keep them for the rest of their lives, regardless of their reproductive or lactating capability.

Monkey Milk

Rhesus macaque monkey mothers produce different milk for their sons and daughters.

They make a milk for their daughters that is less concentrated, but it still provides the same nutrition because they produce more of it. As a result, the daughters have to hang around their mothers for more frequent feedings – and this helps consolidate their matriarchal society.

The sons receive a more concentrated milk, so they spend less time with their mothers and more time playing and exploring. These skills will be useful in their adult lives when they leave the group.

Breast Cancer

Back in the Renaissance, the doctor Bernadino Ramazzini was the first to note that breast cancer occurred more frequently in nuns than other women. Today we know that if a woman delivers and breastfeeds her first child before she is 20, she has about half the risk of breast cancer than either a mother who had her first child after the age of 30 or a woman who has never had a child.

Mama Bear and Cub

There is an old saying in North America,
"Never get between a mama bear and her cub".
Bears will defend their young ferociously – as will
macaque monkeys, rats, mice, prairie voles, hamsters,
lionesses, deer, domestic cats, rabbits, squirrels and
even domestic sheep. In mammals this is called Lactation
Aggression or Maternal Defence.

And according to research by Dr Jennifer Hahn-Holbrook
and colleagues, human mothers can now be added to the
list of Lactating Aggressors. Dr Hahn-Holbrook found that
breastfeeding mothers were about twice as aggressive as
either formula-feeding mothers or women who had never
been pregnant. And just like non-human mammal mothers,
they were aggressive with a cold calm. Their blood pressure
rose less (under the same aggression-inducing circumstances)
than it did for the formula-feeding mothers.

Dr Hahn-Holbrook speculates that in the ancestral
past, this aggression may have successfully deterred
predators. Breastfeeding mothers wouldn't deliberately
place themselves or their infants in harm's way, but
if attacked would defend themselves
vigorously and aggressively.

6

CAN I HEAR SEASHELLS BY THE SEASHORE?

It's a lovely experience to walk on the beach, and it's made even lovelier by finding a large empty seashell, putting it to your ear and hearing the sounds of the ocean. It also gives adults a feeling of benevolent omnipotence to pass the shell to kids and see the amazement on their faces. The ocean can't possibly be inside the shell, so the sounds of the sea coming from the pink walls of a seashell seem like magic.

So what are you actually hearing in the shell?

The answer is that you are hearing the local noises around you, but altered by the shell – thanks to some clever physics. Why do these local noises sound like the sea?

WRONG ANSWER

One popular (but wrong) explanation is that you are listening to your own blood.

This is the explanation given all over the internet (reliable as a drunk person in the pub, etc, etc) and in many books, including *Amazing Facts About Your Body* by Gyles Daubeney Brandreth, published way back in 1981.

> If you do the fancy mathematics, the Shell Near Your Ear acts as a Second-Order Underdamped Lumped Element Low-Pass Acoustic Filter.

This explanation might stem from the fact that you can sometimes hear the pulsing of blood as you lie your head onto a soft pillow.

It's easy to disprove this theory with a simple experiment. Press your ear to a shell and listen, then run around on the beach for a few minutes to increase the blood flow through your body, and listen again to your magic shell.

You'll find that the volume of the "Sound of the Sea" is still the same.

CORRECT ANSWER – IN THREE PARTS

The first part of the explanation is that the shell acts like a "resonator", or a reflector of sound. (You can be less romantic and use an empty coffee cup, or your cupped hand, as a resonator instead.)

When you blow air strongly through your pursed lips over the mouth of an empty bottle, you will hear a musical note as the sound resonates in the bottle. Well, you and I might call it a "bottle" – but a Physical Acoustician would call it a "resonant cavity".

Getting back to our seashell, the inside is hard with an almost glazed finish – so it's an excellent reflector of sound. Most shells also have quite irregular shapes, so they will resonate at many frequencies.

Underwater Sounds

Most low-frequency sounds (20–500 hertz) under the water are caused by distant shipping. There are more ships in the Northern Hemisphere, so there is more low-frequency noise there than in the Southern Hemisphere.

Above that (500–100,000 hertz), underwater sound is mostly caused by the bubbles and spray of breaking waves. It gets louder as the wind speed increases.

Marine creatures such as blue whales, dolphins and harbour porpoises can cover the range from 10–100,000 hertz. For example, blue and fin whales "moan" at 10–25 hertz, while snapping shrimp generate sounds around 2000–5000 hertz.

Physical Acousticians acknowledge that the Ocean in a Seashell is a complicated problem, and that they don't yet have the full answer. They regard the air in the gap between the shell and your ear as the Acoustic Mass. They see the air inside the shell as a separate entity – the Acoustic Capacitance. If you do the fancy mathematics, the Shell Near Your Ear acts as a Second-Order Underdamped Lumped Element Low-Pass Acoustic Filter. This techno-talk just means that it changes (or filters) the ambient sound – similar to how yellow-tinted sunglasses change your visual perception of the outside world. Those sunglasses let through more yellow light, but block blue light.

In the same way, this Acoustic Filter (or Shell Near Your Ear) dampens one frequency (or pitch) and boosts other frequencies.

> If you go into a soundproof room and listen to your favourite seashell you'll hear nothing.

In one study, in a typical noisy room, a cup was held to the ear and a tiny microphone held right next to the eardrum. (The scientists chose a cup because it has a simpler shape than the internal complexities of a seashell.) The sound picked up by the microphone registered 15 decibels louder at the cup's resonant frequency of 648 hertz (compared to not having the cup there at all). But at 1296 hertz, the sound heard was 16 decibels quieter.

But to give you the ocean sound, the shell definitely needs the ambient or background sound. No ambient sound, no Ocean-in-the-Shell sound. If you go into a soundproof room and listen to your favourite seashell you'll hear nothing.

The second part of the explanation is that our human brain is superb at finding subtle patterns in the chaotic world around us. Just like we can find animals in clouds, or the face of Jesus in a potato chip, or the Virgin Mary in a fencepost, we can "hear" the ocean in a seashell.

The third part of the explanation is that we live in a sea of sound, but we mostly ignore it. This is similar to the phenomenon of being able to feel our socks and underwear for a few brief moments after we put them on. Then our brain blocks the socks and underwear from our consciousness for the rest of the day. In the same way, our brain usually blocks most of the noise of the background buzz.

Brain Blocks Sound

You can easily prove that your brain continually
blocks sound. Just go to a hi-fi shop in a big city and
test some noise-cancelling headphones.

These headphones "listen" to the surrounding noise,
make an upside-down copy of it, and then broadcast it
so that both the original sound and the upside-down
version land on your eardrum at the same instant.
The surrounding noise and the upside-down noise
mix and cancel out, generating blessed silence.

Wear these headphones for a minute or two. Then switch
them off (but leave them on your head). Suddenly you
can hear a deep, almost subsonic rumble. This is
the rumble of the city, which reaches many
kilometres past the last suburbs and houses.

And just as suddenly, after less than a minute, you can't
hear it any more. Your brain doesn't see it as a threat
and has "switched it off" from your consciousness.

City dwellers spend their lives drenched in this perpetual
noise – and their brains deliberately ignore it.

TAKE-HOME MESSAGE

So now we can put it all together.

The shell close to your ear acts like the audio equivalent of yellow-tinted sunglasses. It changes the make-up of the sounds that continually assault our ears, and that we continually ignore. To be specific, it lets through more of one frequency, but less of another frequency. So the ear–brain combination recognises that something has changed in the incoming noise. The brain tries to put a label on this new noise, and notices that you are near the ocean – so it labels this noise as "ocean".

But some people find different patterns in seashell noise.

There's a strange para-psychic routine called "Shell Scrying". It encourages you to listen carefully to the shell. First you should hear fragments of words, then words, and finally, whole segments of conversation.

But good luck if you're hoping to hear next week's Lotto numbers . . .

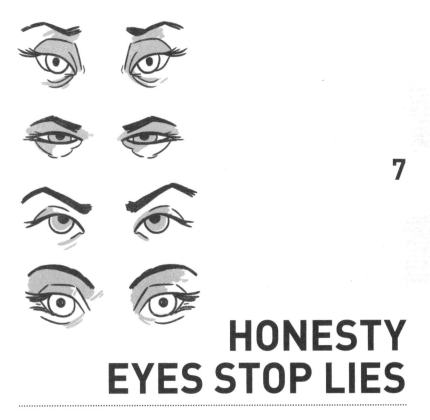

7

HONESTY EYES STOP LIES

To make sure that we stay alive, we humans have evolved a few basic requirements. Hunger and thirst are two of these drives. But if you know a little psychology, you can manipulate this essential need for nourishment to make people more honest.

WHY BE HONEST?

Our society is complicated. If everybody were 100 per cent selfish and totally self-interested, the whole fabric of our society would soon be torn apart. It's absolutely essential to have some degree of cooperation.

Yes, with regard to generosity, there are the wise teachings of the Buddha, Allah, Jesus and other Good Folk. But perhaps part of the reason for generosity is the desire to keep what psychologists call your "pro-social reputation".

But, people *do* tend to be generous, even to those whom they do not know. They will even be generous to people when there is absolutely no chance that they will ever be directly repaid for their generosity.

Over time, people who have built up a history of non-cooperation are shunned, and so they pay a long-term cost for their selfish behaviour. And, sure enough, laboratory studies have shown that people will be more cooperative when they know that their behaviour is being observed.

In fact, it doesn't have to be a real person watching you, or even a surveillance camera.

THE DICTATOR GAME

Consider that game beloved by economists – "the Dictator Game". Player A is the Dictator, and is given some money. Player A can share some of this money with Player B – if they feel like it.

When the Dictator Game was played anonymously (over a computer) so the players couldn't see each other, most Dictators did not share.

But if the computer screen-saver had a pattern that resembled eyes the Dictator was more generous – even if the computer screen carried a background picture of a robot with human-like eyes.

In other words, any image in the local environment that looked like eyes would make the Dictator more generous.

Honest Fish

Eyes watching them can make fish more honest.

A coral reef is a complex ecosystem. Some fish live by eating parasites off other fish. But if these fish know that other fish are watching them, they behave better. They are less likely to bite and eat their customers, and more likely to do a better job of cleaning off the parasites.

TEAROOM EYES – THE SET-UP

In 2006, Dr Bateson and her colleagues from the University of Newcastle in the United Kingdom re-did a version of the Dictator Game in a more realistic setting.

The Division of Psychology had a tearoom. They also had a staff of 48, who could pay for their daily cuppa by voluntarily putting money into an "honesty box". They had been doing this for years. The instructions for how much they should pay were printed on a sheet of A5 paper, displayed at eye height right above the counter. The prices for tea, coffee and milk were 30, 50 and 10 pence, respectively. The counter had everything needed to make the coffee or tea, as well as the honesty box. The fridge containing the milk was right underneath the counter.

Dr Bateson carried out a rather neat experiment on her colleagues in the Division of Psychology. She very subtly altered the instruction sheet carrying the prices for tea, coffee and milk.

For the first week, in a narrow band running across the top of the A5 instruction sheet was an image of a pair of eyes. During the next week the eyes were replaced by flowers. For the third week, a different image of eyes appeared. And so it went on, eyes and flowers, eyes and flowers, for a total of ten weeks. Each week, the pictures of the eyes and the flowers were different. But one thing was constant. The eyes always seemed to be looking directly at you.

This was all done surreptitiously, so none of the other staff in the Division of Psychology realised that the poster carrying the instructions about the payments was subtly changing every week. Dr Bateson made sure that the tea, coffee and milk never ran out. She measured how much milk was used as a reasonable index of how much tea and coffee was consumed.

TEAROOM EYES – THE RESULTS

The outcome was astonishing. In the weeks when the image of the eyes was displayed, the Psychology staff paid 2.76 times as much into their honesty box.

We know that the drive for food and drink is essential for our survival. We also know that our brain has neurons that respond specifically to images of faces and eyes. As Dr Bateson said, "Our brains are programmed to respond to eyes and faces whether we are consciously aware of it or not."

So it's quite reasonable that the images of the eyes in the tearoom led the Psychology staff to believe (subconsciously) that they were being watched. And, in turn, perhaps this motivated the Psychology staff to be more "cooperative" or, in plain English, to pay for the tea and coffee that they were drinking.

So with regard to honesty boxes and fruit stalls on country roads, and even the signs that warn you of a speed camera, maybe the warning signs should go beyond mere words. Maybe they need a picture of the Mona Lisa.

Unconscious Brain

We are always reacting to inputs that our
senses don't consciously register.

Suppose you are asked to look at a photo of a person.
But the experimenters ask you to first take a big whiff of
an odour. The smell, if offered in a high concentration,
might be either pleasant or unpleasant. However, the
odour is delivered to you at such a low concentration
that you can't actually consciously smell it.

Even though you said you couldn't smell anything, you
will more frequently dislike the photo if you have
unwittingly sniffed an offensive odour.

8

CREEPY APP
STALKS WOMEN

The word "app" is short for "application". An app is simply a computer program that has been written to do a specific job. But sadly, Grasshopper, it pains me to tell you that not all Technology is good.

Let's look at the app called "Girls Around Me". It would give you a map with the locations of all women in your area. The women had to be registered with both a location-based social app called Foursquare, and also with another location-based social media application. Foursquare "shouts out" to other Foursquare users that you have checked into a location, while the other social media application carries all kinds of personal information that you and/or your friends have chosen to upload to it (photos, past and upcoming events, etc). The app "Girls Around Me" then used that information – and not for good.

GIRLS AROUND ME

The front screen (or Splash Screen) of "Girls Around Me" gave a pretty good idea of what it was all about. A military-style radar was overlaid on top of a Google map. Holographic green women posing like pole dancers in silhouette would pop provocatively up. And when you pressed the "radar" button, the map would then fill with pictures of women who were in your local neighbourhood – right at that moment. It may or may not have been accidental, but the women were presented as targets. Read into that what you will. After all, it was called "Girls Around Me", not "People Around Me", or even "Women Around Me". Maybe a more honest name would have been "Get the Girl".

On one level, this creepy geolocation-based app was a way to stalk women without their knowledge, and at the same time, get lots of personal and intimate information about them – scary! But on another level, it's a wake-up call for all of us to be more careful about the often-ignored Privacy Settings in the social media that we use.

For example, just by clicking on the icon of a woman in a bar a block or two away, you would instantly see all kinds of personal information about her. Mary-Lou might be single and aged 21, her favourite movie is *Titanic*, she loves blues music, especially Bo Diddley, and you know exactly what schools she went to, and where she went on each of her holidays, as well as the names and birthdays of family and friends – and what the heck, here's a photo of her in a bikini in Tahiti holding her favourite cocktail. This gives a predatory and unethical person an unfair advantage, if they are using this information for nefarious purposes.

PRIVACY SETTINGS: BORING?

You might say that if Mary-Lou was worried about this, she should have paid more attention to the Privacy Settings on her various social networking applications.

It's not that easy, especially when there are more than a hundred options to choose from.

In most cases, the default option is that your Privacy Settings allow everybody access to all of your information. Obviously, it should be the other way around.

> The default option is that your Privacy Settings allow everybody access to all of your information.

And then, when you try to actually adjust the Privacy Settings, you might find it really confusing – because there are so many of them, letting you minutely fine-tune the Privacy Details of each part of your social networking account. And when you finally have learnt how to toggle all the various settings and are comfortable with them, the social networking application changes the options and the wording of the Privacy Settings, and you have to go through a whole new learning process. It's not too hard for technically minded people, but the vast majority of us have to struggle to relearn all the Privacy Settings. So in most cases we do the bare minimum, or nothing.

The app "Girls Around Me" has been pulled from Apple's App Store. But other apps that do essentially the same thing – and more – will pop up. At the very least, we should all learn how to better use the Privacy Settings in our social networking applications. Or we should push for the maximum Privacy Settings to be the default option.

Girls Around Me

This app was developed by the Moscow-based company
i-Free Innovations. The employee who developed
"Girls Around Me" gave an interview to John Brownlee.
The interviewer, John Brownlee, entitled his article
"Our App's Not for Stalking Women, It's for
Avoiding the Ugly Ones".

Joe Bloggs
1946
&
still waiting
for cousin Phil
to arrive...

9

DEATH TAKES A HOLIDAY

When I was a medical student I often heard it said that dying people can deliberately postpone their time of death. According to this commonly held belief, they do this so all the far-flung relatives can arrive in time to say their goodbyes, or so that the dying person can reach a personally important event, such as their birthday or a family celebration or the footy grand final.

Eventually, I found a paper that analysed this belief – and partly debunked it. The researchers rather cutely referred to this as the Death Takes a Holiday Effect.

In their article, entitled "Holidays, Birthdays, and Postponement of Cancer Death", Donn C. Young and Erinn M. Hade analysed every single recorded death from cancer that happened in the state of Ohio between 1989 and 2000 – all 309,221 of them. These 309,221 people were roughly equally divided into male (52 per cent) and female (48 per cent), and were about 89 per cent white and about 99 per cent non-Hispanic. Their median age at death was 72 years, but the range was broad – from one day old to 112 years of age. For the study, the authors chose people who died of cancer because they tend to die slowly – and this might theoretically allow a person to deliberately postpone their own death.

> Looking at Christmas, Thanksgiving and the person's birthday, there was no statistically significant difference in cancer deaths.

The researchers chose to look at three significant dates of the person's life – their birthday, Christmas Day (25 December) and the American holiday of Thanksgiving (the fourth Thursday in November). Christmas Day and Thanksgiving were probably very important to their white subjects in Ohio. They looked at the number of cancer deaths in the week each side of that date.

The results were relatively clear-cut. Overall, looking at Christmas, Thanksgiving and the person's birthday, there was no statistically significant difference in cancer deaths between the week before that date and the week after that date.

However, in direct opposition to their findings, there were some previous studies that claimed to have found some truth in this Death Takes a Holiday Effect. This discrepancy bothered Drs Young and Hade, so they went back and analysed those earlier studies.

First, they found that many of those studies had sample sizes that were too small to be able to draw any reliable conclusions. Second, some of these studies "massaged" the small pool of data over and over until they found a result. For example, one study performed over 100 statistical tests to a sample size of only some 400 people.

Maybe Drs Young and Hade themselves should do a bigger study. After all, they looked at cancer deaths in Ohio from 1989–2000 – which was only one-third of all deaths. I can't help but wonder if they would have gotten different results if they looked at all deaths in Ohio in that period.

The Indian poet, playwright and essayist Rabindranath Tagore, who won the 1913 Nobel Prize for Literature, wrote, "Death is not extinguishing the light; it is only putting out the lamp because the dawn has come."

Selective Memory

So why do we humans tend to see patterns in our lives that don't stand up to close analysis?

First, the Universe is Big and Complicated, and we have to simplify it to make sense of it all.

Second, as part of this simplification, it turns out that our brains are prone to Selective Memory. The Death Takes a Holiday Effect is a great example of these patterns.

We easily remember events that happened on "significant" days, such as Christmas Day, while events that happened on random days tend to get lost in our memories.

Letting Go

So, it seems that people dying of cancer usually can't hang on for a special occasion. But it appears that the opposite might be true – that people can "give up" or "let go".

This is simply personal speculation – I have found no paper on this topic. It is purely based on stories from people in the health profession. They often report that if the patient is near death, they can say goodbye to friends and family and then just die. Sometimes they say they are "tired of fighting".

10

ALCOHOL AND DEHYDRATION

We humans have been making and drinking alcohol for thousands of years. This intoxicating liquid seems to be able to do anything. We use it as fuel for cars and to kill germs, it can preserve human heads or other body parts in jars for years on end, and it will even strip oil stains from the garage floor. And yet, in small quantities we also use alcohol as a social lubricant.

All drugs are poisons, but what matters is the dose. Over time, too much alcohol can result in numerous health problems including diabetes and malnutrition, and diseases of the central nervous system and the liver. A short-term side effect is excessive urination. Shakespeare knew this, and so, in his play *Macbeth*, the porter says that alcohol promotes "nosepainting, sleep and urine".

But even today we still don't fully understand how alcohol causes this excessive urination.

BEER TO URINE

After all, the average beer is about 95 per cent water and only 5 per cent alcohol. And the liver converts that 5 per cent of alcohol into roughly the same mass of water and some carbon dioxide. So if you drink 200 millilitres of beer, the end result is near to 200 millilitres of water.

> If you drink 200 millilitres of beer, you urinate a total of about 320 millilitres.

But you don't urinate just 200 millilitres of urine after drinking 200 millilitres of beer – no, you urinate a total of about 320 millilitres.

So in general, each shot of alcohol makes you urinate an extra 120 millilitres on top of your normal urine output. Where does that extra 120 millilitres come from?

To understand what's going on, you need a bit of background knowledge.

ALCOHOL AND YOUR BODY

First, the body pays special attention to alcohol, as compared to many other small molecules. So alcohol gets carried very quickly through the walls of the gut into the bloodstream and then to the brain.

Second, for some unknown reason, we humans seem to prefer to drink our alcohol in 10-gram "lumps". A Standard Glass of beer, wine, or spirits contains about 10 grams of alcohol.

Third, alcohol interferes with the mechanism that regulates the water levels in our body. At this stage, I need to explain a little anatomy and physiology. In your brain is a small gland called the pituitary gland. It is divided into two sections – the front and the back. The back section is called the posterior pituitary. One of the hormones made by the posterior pituitary gland is vasopressin, or Anti-Diuretic Hormone (ADH). "Diuretic" is a fancy word related to "urination". The job of ADH is to stop you urinating.

Urinate More in Winter?

If you weigh 60 kilograms, you generate about 60 millilitres of urine each hour (and for 80 kilograms, about 80 millilitres per hour, and so on).

In winter, your skin is colder, and so more blood is diverted from your skin to your kidneys. More blood going into the kidneys means more urine coming out.

Women in late pregnancy already have to urinate more frequently because the baby is pressing on the bladder. Adding winter to the equation means lots more visits to the bathroom.

DEHYDRATION VERSUS ALCOHOL

So how does ADH work? Let me explain.

Suppose that you are really dehydrated. This means the volume of water in your body is low. But you still have just as many salts floating in this reduced volume of water. So these salts are now more concentrated.

Your body has detectors that can sense both the saltiness of your water and the volume of that water. If these detectors reckon that you are dehydrated, they send a signal to the posterior pituitary gland, which starts pumping out ADH. ADH tells the kidneys to slow down your rate of production of urine – so you hang on to your precious water. You reduce your normal rate of making urine.

Alcohol opposes this action. It acts directly on the posterior pituitary to reduce how much ADH you make. Less ADH means that you make more urine. Each standard drink of alcohol that you drink forces your kidneys to generate an extra 120 millilitres of urine on top of the average 60–80 millilitres per hour that your body usually makes.

Already Dehydrated?

What happens if you are already dehydrated, and then drink alcohol?

The body is "clever" enough to tell the kidneys to reduce the excess urination that normally happens after drinking alcohol. So you will generate more urine than if you had not drunk any alcohol at all, but less than if you started off normally hydrated.

DRINK MORE WATER?

Aha, you cleverly think to yourself, why don't I just drink lots of water to compensate for the extra 120 millilitres urinated? Unfortunately, it's not that simple.

You'll hang on to only about half or a third of the extra water you drink. Most of it will go out in your urine, and you'll still end up dehydrated at the end of a night of drinking. Mind you, you'll be better off than if you didn't drink any extra water at all – but you'll still be dehydrated.

So this extra urination (caused by alcohol consumption) is perfectly consistent with that old Aussie expression, "a night on the piss".

11

FAST FOR ONE YEAR: THE MAN WHO ATE HIMSELF

Back in June 1965, a "grossly obese" Scotsman weighing 207 kilograms, and hereafter known only as Mr A.B., turned up at the Department of Medicine at the Royal Infirmary in Dundee. He was sick of being fat and wanted to lose weight. He told the hospital staff he was going to fast, whatever they said, so they might as well monitor him along the way. He ended up fasting for one year and 17 days – that's right, he ate no food at all for over a year. He lived entirely off his copious body fat, losing about 125 kilograms of weight.

The first thing I should point out is that treating obesity by total starvation can be dangerous. There are many reports of total starvation leading to death. For example, people have died of heart failure while fasting. Another person died on the thirteenth day of his fast from a small bowel obstruction. Others have died during the refeeding period after the fast – one from lactic acidosis.

But on the other hand, going hungry is natural (but starvation takes hunger to extremes). *Homo sapiens* evolved over the last 200,000 years and for most of that time, food was not always freely available. We have evolved to deal with periods of not enough food. In fact, there are some studies showing that fasting (or, at least, calorie restriction) can have health benefits under certain circumstances.

> **Treating obesity by total starvation can be dangerous.**

Epilepsy Benefits of Fasting

Surprisingly, fasting can help treat epilepsy. Galen wrote that the ancient Greek royal physician Erasistratus said, "One inclining to epilepsy should be made to fast without mercy and be put on short rations".

Thanks to modern biochemistry, we know that fasting leads to the breakdown of fat, which in turn leads to ketones being produced throughout the body. There are modern strict ketogenic diets designed for children with epilepsy that are resistant to multi-drug treatment. Do they work? Well, about one third improve dramatically, one third get some improvement and the remaining third get no benefit.

This ketogenic diet has very little carbohydrate, moderate amounts of protein, but lots of fat. This diet forces the neurons in the brain to change from their regular fuel of glucose to ketones (a byproduct when fats are broken down). It is thought that this happens via a protein with the unfortunate name of BAD (BCL-2-Associated Agonist of Cell Death).

The benefits of fasting as a treatment for epilepsy were rediscovered in France in 1911, and in the USA in 1916. Fasting fell out of favour in the 1930s when anti-epileptic drugs were developed. But Johns Hopkins Hospital in Baltimore, Maryland, has never stopped using it.

While it definitely has benefits for some children, there are side effects. Kidney stones develop in one in every 20 children on this ketogenic diet (as compared to one in several thousand for the general population). Cholesterol levels increase by 30 per cent, puberty can be delayed, and there can be both retarded growth and an increased risk of bone fractures.

This treatment was somewhat popularised by the 1997 made-for-television movie called, . . . *First Do No Harm*. It starred Meryl Streep. And yes, in the movie a ketogenic diet successfully treated a young boy's previously untreatable epilepsy.

STARVATION RESPONSE

We humans invented agriculture some 10,000 years ago, and ever since, our food supply has become more secure and reliable. Today, most people in developed countries eat three meals per day, often with snacks in between.

So what fuels your body?

For those of you who have not studied biochemistry, here's the short version. Your body gets its energy either from glucose (when you are eating) or ketones (when you are not eating, but are breaking down fats into ketones).

Now for the longer version of what happens when you don't eat.

There is a transition between eating and not eating after your last meal. Your body still gets energy from the glucose in your bloodstream and liver. You carry a semi-permanent 0.5–1 kilogram of solids in your gut. The glucose from this runs out after about 8 hours.

Then you start burning up a chemical called glycogen. Glycogen is simply a whole bunch of glucose molecules loosely stuck together. It's stored in your liver and muscles. Glycogen is really easy to break down into the individual glucose molecules from which it was made. You can burn glycogen to get the glucose you need for about another 36–48 hours.

After two or three days of fasting, you are getting your energy from two different sources simultaneously. A very small part of your energy comes from breaking down your muscles – but you can avoid this by doing some muscle work (Resistance Training, otherwise known as Pumping Iron). However, the majority of your energy comes from breaking down fat (long-chain triglycerides).

But very soon you move into getting all your energy from the breakdown of fat. The fat molecules break down into two separate chemicals – glycerol, which can be converted into glucose, and free fatty acids, which can be converted into other chemicals called ketones. Your body, including your brain, can run on glucose and ketones until you finally run out of fat. (Just as an aside, microbes have used ketones for energy for well over 3 billion years.)

> Your body gets its energy either from glucose (when you are eating) or ketones (when you are not eating).

The average, non-obese 70-kilogram male carries about 8000 kilojoules of energy in glycogen, and about 400,000 kilojoules in his body fat.

If you don't start eating once you run out of fat, your body will move on to desperate measures. It will begin to break down your muscles to get the energy it needs. Each gram of protein will produce about half a gram of glucose. Death will not be far away. Arctic and Antarctic explorers who tried to live entirely on meat without eating blubber have learnt this lesson the hard way.

Blood-pressure Benefits of Fasting

One study looked at 174 people with high blood pressure. They fasted for ten days, by which time blood pressure had returned to normal for 154 of them. Averaged out over all patients, the drop in systolic/diastolic blood pressure was 37mmHg and 13mmHg respectively. In the subgroup of the patients with the most severe high blood pressure, the average drop in systolic pressure was 60 mmHg.

The author of the study, Dr Alan Goldhamer, claims that this is the greatest reported drop in blood pressure achieved by any drug or therapy. He says that in 42 patients whom he monitored six months after the fast, blood pressure had remained at its new low level.

I really wish that he had monitored all of them, not just one third.

FAST FOR OVER A YEAR

In the case of our Scotsman, Mr A.B., the staff in the Medical School at the University of Dundee kept a close eye on him. He did not eat any food, but the staff gave him yeast for the first 10 months and multivitamins every day of the fast. Potassium is essential for the proper working of the heart, and when his potassium levels got a little low, he was given effervescent potassium tablets from Day 93 to Day 162. He defecated infrequently, roughly every 37–48 days.

Blood samples were taken every fortnight, and his carbohydrate metabolism was checked on nine occasions during the 382 days of his fast. Surprisingly, for the last eight months of his fast, his blood glucose levels were consistently very low. They were around 2 millimoles, which is about half of the bottom end of the normal range (3.6–5.8 millimoles). Even so, he did not suffer clinically from this abnormally low blood glucose level. (Not all "therapeutic fasters" have had low blood glucose levels recorded.)

According to the *Guinness Book Of Records*, his is the longest recorded fast. This record has since been removed from the publication to stop unsupervised fasters from trying to break his record, and possibly harming themselves. His weight dropped from 206.8 kilograms to 81.6 kilograms. Some five years later, he had regained only 7 kilograms.

Babies

Humans are the only primate usually born facing backwards – and the only primate born obese.

Newborn babies have very low blood glucose levels, and high levels of ketones. The brain of a newborn baby uses up about two thirds of all its energy – and about half of this energy comes from ketones. (In adults, the brain uses up about 20 per cent of our energy, even though it weighs only 2 per cent of our body weight – and it usually runs entirely on glucose.)

Babies start life on what is basically an Atkins Diet. Their mothers' colostrum is rich in triglyceride and protein, but very low in lactose. Over the next two to three days, the lactose level in the breast milk increases and the babies' ketosis disappears.

The History of Fasting

Ancient thinkers from Greece and Rome, such as Socrates, Plato, Hippocrates, Aristotle and Galen, recommended that a short fast would clear both the mind and the body. The Greek historian Plutarch advised, "Instead of employing medicines, fast for a day".

The Greek mathematician Pythagoras – the right–angled triangle dude – applied to study in Egypt. Before he could be accepted, it is said that he had to fast for 40 days. He did so unhappily at first, but at the end of 40 days declared that he was reborn. In later life, he would force his students to undertake fasts.

FASTING – PROS AND CONS

Fasting seems to have health benefits for high blood pressure, epilepsy in children, diabetes and asthma – in humans. In animals, it seems to reduce the cognitive decline that happens in illnesses such as Parkinson's disease and Alzheimer's disease.

One problem with fasting is that you can become cranky and irritable (and sometimes hungry). Another is the lack of companionship with your fellow human. After all, the word "companion" comes from the roots "com" meaning "with" and "panis" meaning "bread". If you spend lots of time not sharing meals with your kith and kin, you run the risk of becoming an outsider.

19th-century Fasting

On 17 July 1877, a Minneapolis doctor called Henry S. Tanner decided to commit suicide. He was not a happy man. He was generally physically unwell for no specific reasons, had not been a great success in either his ownership of a Turkish bathhouse or being a temperance lecturer, and his wife had left him. So on the fateful day he had decided he would do himself in, he drank a pint of milk and went to bed to melodramatically starve himself to death. After all, the then-current scientific opinion was that a human would die after ten days without food.

But it all went terribly wrong!

By the tenth day, his non-specific illness had vanished, along with his hunger, and he was not near death. He felt renewed vigour and strength coursing through his body. Many years earlier, before his life went belly-up, he used to walk up to 5 kilometres twice a day. On the tenth day, he leapt from bed and began the first of many long walks. On the 41st day, he finished his fast with a glass of milk.

In 1880, Dr Tanner went on a 40-day fast on the stage of the Clarendon Hall on East 13th Street in New York. The public were invited to observe him sitting on the sparse stage, furnished with only his cot, a cane-backed rocker and a gas-fired chandelier. He did absolutely nothing – apart from drink water, read and chat with the spectators, who had paid the princely sum of 25 cents for the privilege.

Thanks to newspaper publicity across the USA, he began to receive exercise equipment, flowers, wine, slippers, mattresses and up to 400 letters a day. A woman in Philadelphia proposed to marry him if he were to survive, while the director of a museum in Maine offered to stuff and display him in his museum should he die.

He finished his fast on the 40th day by eating a peach, having dropped his weight from 71.4 kilograms to 55.1 kilograms. He then drank two goblets of milk and ate most of a watermelon, which he soon followed with half a kilogram of meat, some apples, and wine and ale. By the next evening he had regained 3.9 kilograms, by the third day 8.8 kilograms, and by the eighth day he had regained all of his lost 16.3 kilograms.

12

PLANET K: 18412 KRUSZELNICKI

Richard Branson and I have shared a stage and a Green Room –
though he might not remember it as clearly as I do. From what I saw
in our brief encounters, he seems to have his heart in the right place.
He has also worked hard and been lucky, and has saved up enough
pocket money to buy his own island in the Caribbean.

Well, he might have his own island, but I have my own minor
planet. Yes, that's right, I have a planet named after me.

MY OWN MINOR PLANET

When I say that a planet is named after me, I'm not saying that Jupiter or Saturn have been renamed "Dr Karl".

No, my planet is rather minor – an asteroid, actually. It's not thousands of kilometres across – more like 25–100 metres across. It used to be called 1993 LX, but now it has been renamed 18412 Kruszelnicki (1993 LX). New minor planets are always been discovered. "My" minor planet, 18412, was discovered by R.H. (Robert) McNaught back on 3 June 1993.

> My planet, 18412 Kruszelnicki, is in a roughly circular orbit in the asteroid belt, between Mars and Jupiter.

My planet, 18412 Kruszelnicki, is in a roughly circular orbit in the asteroid belt, between Mars and Jupiter. An Astronomical Unit (AU), the distance between the Earth and the Sun, is about 150 million kilometres. The orbit of Mars ranges between 1.4 and 1.7 AU, while Jupiter's varies between 5.0 and 5.6 AU. "My" little rock orbits between 2.8 and 3.3 AU and takes 5.3 years to loop around the Sun.

This is a genuine naming. This is not your Pay $10 To Own A Star or Pay $15 To Own A Square Metre Of The Moon. Nope, this is all genuine and above board. It's registered with the International Astronomical Union.

I don't actually own the minor planet, or have any mining rights, etc. My name was linked to 1993 LX as a courtesy by the discoverer, Robert McNaught.

On 17 April 2012, Robert McNaught sent me an email which read:

Following the suggestion of Steve Quirk, an amateur astronomer, avid radio listener and science buff, I have named a couple of asteroids in honour of Karl and Adam. The asteroid numbers and official (International Astronomical Union) names are given at the end of the message, along with the official naming citation.

The "Adam" referred to is Adam Spencer, my fellow Sleek Geek in our various TV series and roadshows around Australia. The email went on:

A JPL website can be used to visualise the orbit of (18412) Kruszelnicki: *http://ssd.jpl.nasa.gov/sbdb.cgi?sstr=18412;orb=1;cov=0;log=0;cad=0#orb*

The page also displays all the known orbital and physical data on the object.

For (18413) Adamspencer, go to *http://ssd.jpl.nasa.gov/sbdb.cgi?sstr=18413;orb=1;cov=0;log=0;cad=0#orb*

Please enjoy! I suspect there will be rules regarding real estate development, mining, or deflecting the object onto rival's [sic] asteroids," but hey, just go for it.

And that's how I got "my" own minor planet.

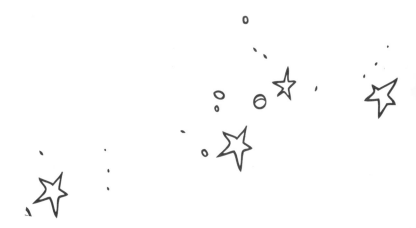

Robert McNaught

Robert is an astronomer who gave many hours of
pleasure to lots of people after he discovered
the comet C/2006 P1 on 7 August 2006.

This comet was renamed Comet McNaught in his honour.
Because it was the brightest comet for four decades, it was
also given the name of The Great Comet of 2007. After its point
of closest approach to the sun on 12 January 2007 and for the
next few days, it could be seen around the world in daytime,
about 7 degrees southeast of the Sun. My family had many
hours of great pleasure watching it at twilight with the
naked eye in January and February 2007. It was so
much fun to point it out to random passers-by.

Robert McNaught is at the Research School of Astronomy
and Astrophysics at the Australian National University.
Besides discovering over 35 comets, he has also
discovered (as of May 2012) some 410 asteroids.

Mining Asteroids

There are over 20,000 asteroids bigger than 100 metres across in the Asteroid Belt between Mars and Jupiter. Even an asteroid with a diameter of 1.6 kilometres holds about US$20 trillion worth of various metals.

On Earth, when the planet was entirely liquid and molten in the Early Days, the heavy precious metals mostly sank to the centre. Most of these metals we can find today came from the Bombardment of Asteroids that hit the Earth after the crust cooled. We expect to be able to find these metals (such as gold, platinum, palladium, rhodium, etc) still in asteroids.

In April 2012, the start-up company Planetary Resource announced its desire to mine asteroids. It is backed by explorer and Hollywood filmmaker James Cameron, billionaire Charles Simonyi and others.

13

CANCER CREEP

When it comes to cancer, Joe Jackson's song holds out little hope: "Everything Gives You Cancer". And, sure enough, the cancer rate has increased threefold over the last century. So, does everything give you cancer? No, it's just that we live so much longer nowadays.

TWO CENTURIES OF DISEASES

In 2012, *The New England Journal of Medicine* celebrated its 200th year of existence. Over those two centuries, the nature of the diseases that affect us has changed enormously.

In 1812, people died of diseases such as "gunshot wounds", "drinking cold water", "mortification", "teething" and even "being missed by a cannon ball". In those times before antibiotics, there was a whole bunch of fatal diseases – including a strange one called "spotted fever", which caused neither spots nor fever.

> In 1812, people died of diseases such as "gunshot wounds", "drinking cold water", "mortification", "teething" and even "being missed by a cannon ball".

A century later, the pattern of disease had changed again. By 1900, tuberculosis and pneumonia each killed slightly more people than cancer and heart disease do today – around 200 deaths per 100,000 people each year. Mind you, there was also a disease called "automobile knee".

And almost another century later again in 2010, there was a different pattern to the diseases that killed us.

One very surprising finding is that in the USA the "previously steady increase in life expectancy has stalled, and may even be reversed". If this apparent reverse turns out to be real, the current generation in the USA will be the first to have a lower life expectancy than their parents'.

21st CENTURY DEATH

The first change from 1900 to 2010 was that the overall annual death rate had dropped – from 1100 deaths per 100,000 people per year down to 600.

The second change was the increase in heart disease by about 50 per cent. This is partly related to the obesity epidemic in western countries, which in turn is related to overeating and lack of exercise. In the US, 36 per cent of people are obese, and over 70 per cent are overweight. This is partly caused by the fact that many of today's jobs are so-called "thinking" jobs, where we sit and think instead of doing physical exercise.

The third change was a massive drop in deaths due to better treatment and prevention of certain types of diseases. These include infectious diseases such as pneumonia, influenza, tuberculosis and some gastrointestinal diseases. Vaccination eradicated smallpox, and may soon eradicate polio. Genetic screening also led to major reductions in thalassemia and Tay-Sachs disease.

A fourth change is that women now live longer than men. On one hand there were huge changes in antenatal care and obstetric practice, reducing the number of women who die in childbirth, or from pregnancy- or childbirth-related problems (for example, pre-eclampsia). On the other hand, the epidemic of heart disease that peaked in the US in the 1960s affected men more than women.

The fifth change was the tripling of cancer deaths.

This is where it gets complicated. About 5–10 per cent of cancers are caused by genetics. So the remaining 90–95 per cent of cancers are caused by environmental factors – diet and obesity (30–35 per cent), tobacco (25–30 per cent), infections (15–20 per cent), radiation (5–10 per cent) and "other" causes.

So doesn't that mean "everything gives you cancer"?

Yes and No. Yes, a small number (but not all) of environmental factors can lead to cancers. And, No – you usually have to live for a long time to get most cancers. About 80 per cent of men have prostate cancer by the age of 80 – but in most cases, the prostate cancer does not kill them. About 35 per cent of people who die have a thyroid cancer – but the thyroid cancer did not kill them. Most people who are diagnosed with cancer are already over the age of 65.

So, the increase in the cancer rate is overwhelmingly due to the fact that we live longer. These extra years of life allow extra time for the cancer not only to start, but to grow and eventually kill us. As cancer scientist Robert A. Weinberg said, "If we lived long enough, sooner or later we all would get cancer."

The concept of "disease" is complicated. It's important to realise that it's not just doctors or patients who define a disease. As Dr Jones and colleagues wrote in *The New England Journal of Medicine*, "Diseases can never be reduced to molecular pathways" – they're more than that.

Diseases are defined by the social, economic and political processes that shape society itself, including deliberate advocacy by interested parties. So at different times in different societies, the following have been – and not been – considered diseases: homosexuality, alcoholism, masturbation, chronic fatigue syndrome and sick building syndrome.

In one sense, who gets what disease "lays bare society's structures of wealth and power". In a way, poverty itself is a disease.

How Do New Diseases Emerge?

New diseases have popped up over
the last few centuries – and will keep doing so.

Every so often this results from new causes of illness,
such as radiation poisoning or motor vehicle accidents.
Another pathway is new behaviours, such as cigarette
smoking or intravenous drug use.

In some cases, disease treatments can alter the pathway of the
original disease. For example, if diabetics live longer because
of better treatments, there is more time for complications
such as diabetic retinopathy to appear.

Occasionally a once-rare or uncommon disease (for example,
lung cancer or Mad Cow disease) can quickly increase in
incidence. This can happen via changes in environmental
or social conditions.

Sometimes "new" diseases are not really new – they are
actually old diseases that we have recently learnt to diagnose.
A good example is high blood pressure, which has very
few external signs or symptoms.

14

CARBON CAR: THE EMISSION MISSION

There's a story going around that modern, efficient cars use up more carbon in their manufacture than is ever recovered by their greater fuel economy.

Who knows where this story started? But there's an Old Legal Saying, "Cui Bono?" or "Who Benefits?"

CARBON NUMBERS

With a car, there are two main sources of carbon emissions. There is how much carbon you have to burn to produce that car in a factory, and then there is how much carbon comes out the exhaust pipe over the running life of that car – typically about 150,000 kilometres. Finally, "Whole of Life emissions" are obtained by adding "production emissions" and "running emissions".

The good thing about burning carbon is that when it combines with oxygen, one of the byproducts is heat energy. This heat energy can be used to propel the car down the road. (The bad thing about burning carbon is that when it combines with oxygen, the other byproduct is carbon dioxide.)

Emissions (tonnes of CO_2)

	Whole of Life	Running	Production
Petrol Car	24	18.4	5.6
Hybrid Car	21	14.5	6.5
Plug-in Hybrid Car	19	12.3	6.7
Battery Electric Car	19	10.2	8.8

Let's compare a standard petrol vehicle to three other cars – the hybrid, the plug-in hybrid, and the battery electric car. As you move from the petrol car to the battery electric, it takes more carbon dioxide to actually manufacture the vehicle. But this is more than compensated for by the lesser amounts of carbon dioxide needed to move the car down the road.

So when you look at the Whole of Life carbon dioxide emissions, you drop from 24 tonnes for a standard petrol car down to 19 tonnes for a battery electric car.

EVEN MORE EFFICIENT

There are a few more points to consider.

First, at the moment hybrid and battery electric cars are made in relatively small numbers. As the production numbers increase, Economies of Scale will make their manufacture more efficient. Furthermore, both Toyota and Nissan have announced they are already using renewable power (non-fossil fuel) at their manufacturing plants to reduce the carbon emissions associated with producing cars.

Second, if the plug-in hybrid car and battery electric car get their electricity from renewable power, this will further reduce their Whole of Life carbon footprints.

So the Urban Myth that hybrid and battery electric cars are worse in terms of carbon emissions is just that – a myth. If a car is more efficient, then everybody benefits – except maybe the oil companies.

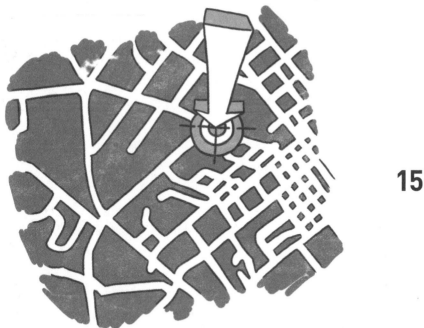

15

GEO-TAGGING PHOTOS: LOCATION, LOCATION, LOCATION

If you don't understand the technology you use, every now and then your ignorance will come back to bite you.

One expensive example of this happened back in 2007 on a US military base in Iraq. Because some soldiers took photographs, the US taxpayer lost about $150 million.

The new-ish technology is called "geo-tagging". What it means is that when you take a photograph or video, your geographical location is almost always embedded in the digital files of that photograph or video. This happens automatically with practically all smartphones that have a camera built in, and with a few specific digital cameras.

There are different ways that your smartphone can know your location. It can use "real" GPS that gets location and time information from GPS satellites orbiting some 20,000 kilometres overhead. Or it can work out your location by knowing the positions of the nearest mobile phone towers, or any WiFi hotspots, including the one in your house.

But regardless of how your smartphone knows where you are, the important thing to realise is that in the vast majority of cases, it will automatically embed your location into every photograph you take. If you want to stop it from adding your whereabouts to your photographs, you have to dive deep into the settings of your smartphone. However, if you do accidentally imprint geo-tags on your photos when you take them, you can later remove them with the many so-called "metadata removal tools" available on the web.

> The important thing to realise is that in the vast majority of cases, your smartphone will automatically embed your location into every photograph you take with it.

This has . . . ahh . . . interesting privacy implications. If you take a photograph of a painting on your bedroom wall, and then post that photograph on a location-based social media application, anyone can work out the location of your bedroom. Another scenario is that a thief can work out that you have gone on holidays by checking your location via the photos you post on the web.

Getting back to the snap-happy soldiers who cost the US taxpayer lots of money: in 2007 they took delivery of some AH-64 Apache helicopters at a military base in Iraq. These are not cheap choppers. In 2012, the cost of a brand-new one was US$38 million. The soldiers proudly took pictures of their new attack helicopters and uploaded them onto the web. Almost certainly, the soldiers did not realise that geo-location tags were embedded in those photographs.

According to Steve Warren, a US Army Maneuver Center of Excellence intelligence officer, the enemy was therefore able to determine the exact location of the helicopters inside the compound and conduct a mortar attack, which destroyed four of the AH-64 Apaches.

That was a $150 million mistake – but I guess that the soldiers now know what geo-tagging is. I wonder how long it will be before we civilians learn about the OMGs of geo-tagging.

Geo-tagging

Your geographical identification data is usually stored as "metadata" in what is called an EXIF file, as part of your digital photograph.

EXIF stands for Exchangeable Image File Format. It is a standard that specifies the format for images and sound used by digital cameras. In addition to storing images and sound, it also stores date and time information, the camera settings that were used, a small thumbnail picture to preview the image on the camera's LCD screen, and so on.

But EXIF files are now being used to store your latitude and longitude coordinates. Furthermore, the geo-data stored can also include your altitude, the direction the camera was facing when it took the photograph, how accurate your camera's estimate of your location is, and even place names.

16

GRANDER CANYONS

In 1903, Theodore Roosevelt said that if American citizens were to see just one natural wonder in their lives, they should visit the Grand Canyon in Arizona. In the Grand Canyon, the Colorado River has carved a unique and beautiful glimpse into the geology of the last 2 billion years. But even though most people think of it as the largest canyon in the world, it isn't. It has massiveness, scenic beauty and stupendous three-dimensional exposure – but it isn't the world's largest canyon.

Nope, the Capertree Valley in Australia is wider than the American Grand Canyon. But it's not as spectacular, and, because of Australia's antiquity, not as deep. It's located about 135 kilometres north-west of Sydney, between Lithgow and Mudgee.

But the deepest canyon on Earth? That's the Grand Canyon of Yarlung, in the south-eastern corner of Tibet.

CANYONS 101

Ditches in the ground with very steep walls are called "canyons" in the USA (from the Spanish "cañón"), or "gorges" in Europe and Australasia.

The Colorado River rises in Colorado, and then flows mostly southwest some 2330 kilometres into the Gulf of California. Along its length, the 446-kilometre-long Grand Canyon plunges over 1.6 kilometres below the rim down to the riverbed. At its widest, it's about 19 kilometres across. It's truly spectacular.

Shangri-la

James Hilton's 1933 book, *Lost Horizon*, places the earthly paradise of Shangri-la in western Tibet. But those "in the know" say it's really located near the awesome Grand Canyon of Yarlung in eastern Tibet. It's accessible (for the enlightened) only via a hole in the rocks, behind the liquid thunder of one of the river's impassable waterfalls.

The Anasazi Indians moved into the Grand Canyon area about 3000 years ago. Spanish explorers saw it in 1540, while Major John Wesley Powell first carefully documented it in 1869.

Now for the question of how to compare one scenic eroded monument with another – do we compare length or depth? The Great Canyon of Yarlung wins on both counts. It's a little bit longer – 496 kilometres long. But it's an astonishing 3.7 kilometres deeper. Yup, it's 5.3 kilometres from the rim to the riverbed.

The Grand Canyon and Dinosaurs

The standard answer to the "How old is the Grand Canyon?" question used to be, "About 6 million years". That was when the surrounding plateau began rising from sea level to its current elevation of over 2 kilometres – and when the then-Colorado River started grinding its way through the rock.

The latest data (although still a bit controversial) suggests a different scenario. In this theory, the Grand Canyon could be 15 million years old, or even might have had its beginnings before the dinosaurs died out, some 65 million years ago.

In the new scenario, geological forces pushed up the Colorado Plateau some 80 million years ago. Back then a river eroded away a 1-kilometre-deep canyon at what is now the eastern third of the Grand Canyon. Then many different rivers sliced away the top two kilometres of the whole Colorado Plateau. It seems that the "original" Colorado River ran in the opposite direction, way back then.

GRAND CANYON OF YARLUNG

This canyon was carved out by one of the mightiest rivers in Asia – the Brahmaputra. Along its 2900-kilometre length, this river has many names – the Yarlung Tsangpo in Tibet, the Siang in India's Arunachal Pradesh, and so on. It eventually meets up with the Ganges, and together they form the largest river delta in the world, in the Bay of Bengal in Bangladesh. In the rainy season, the Brahmaputra carries water at the rate of about 14,200 cubic metres each second.

The Brahmaputra starts in far western Tibet, rising from the Chemayungdung Glacier in the Kailash mountain range. It then flows mostly east for about 1100 kilometres, staying a little north of, and parallel to, the Himalayan mountain range. In the far eastern end of Tibet, it then performs a bizarre 260-degree turn while cutting through the highest mountain chain on Earth, the Himalayas. Here, at the Great Bend of the Yarlung Tsampo, it bends first to the north, then to the east, then the south. While heading north it cuts between two mountain peaks just 20 kilometres apart – the Gyala Peri mountain (7.15 kilometres above sea level, but estimates vary) and Namcha Barwa (7.76 kilometres).

This is the location of the world's deepest canyon. Like the river that runs through it, the canyon goes under many names – the Great Canyon of Yarlung, the Namcha Barwa Gorge, the Yarlung Tsangpo Canyon, the Tsangpo Gorge, etc. There is a very steep drop of over 5 kilometres from the top of the Gyala Peri mountain to the river. The walls of the canyon are three times steeper than the Grand Canyon, while the gradient of the river is eight times greater than the Colorado River's.

The area is very difficult to get to – ranging from snow and glaciers at high altitudes, to impenetrable rainforest at lower altitudes. It also suffers continual landslips and earthquakes, thanks to being at one of the thrust pivot points at which India (travelling north at 5–10 centimetres per year) rams into Asia. The canyon was measured accurately for the first time in the 1990s.

In 1994, the American Geography Committee named it "The Grandest Canyon On Earth", snatching the honour away from the Grand Canyon in Arizona.

EVOLUTION AND CREATIONISM

But recently there have been strange happenings in Arizona. Tom Vail and his wife, Paula, operate Canyon Ministries, which runs "Christ-Centered Motorized White-Water Rafting Trips" in the

Grand Canyon. Tom Vail's book, *The Grand Canyon: A Different View*, argues from a biblical point of view, and fancifully claims that the Grand Canyon is only a few thousand years old, and that it was formed in a few days by massive erosion from the release of huge lakes of trapped water left over after the Great Flood of Noah.

His book has an anti-evolutionary point of view. But it is sold in shops in the National Park that, in turn, fund research into the evolutionary origins of the Grand Canyon. God moves in mysterious ways . . .

Moving Dirt

The Colorado River used to carry 500,000 tonnes of sand, silt and clay downstream every day. That was about 6 tonnes each second. But in 1963 the Glen Canyon Dam was built upstream of the Grand Canyon, so the flow and sediment transport dropped considerably.

Worldwide, all the sediment discharged into the oceans comes to about 10–20 billion tonnes per year. This creates vast "aprons" or "fans", beginning where the river meets the ocean. The Indus Fan and the Bengal Fan are to the west and east respectively of India. They are a few kilometres thick at the river mouth, and thin out gradually seawards over thousands of kilometres.

We humans currently shift about 7 billion tonnes of coal and 2.3 billion tonnes of iron ore each year. One single engineering project, the Syncrude Mine in the Athabasca Sands in Canada, involves moving 30 billion tonnes of earth over the life of the project.

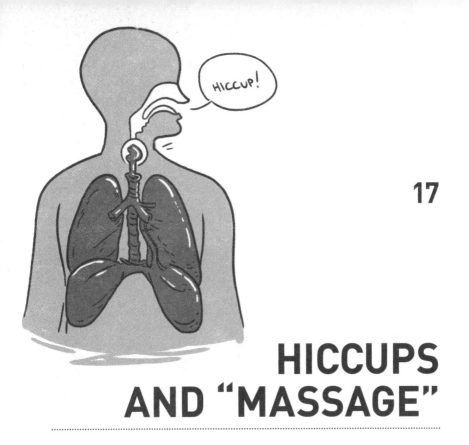

17

HICCUPS
AND "MASSAGE"

You almost certainly had your first hiccup before you were born, probably around 11 weeks after you were conceived. It turns out hiccups are essential to the normal lung development of unborn babies. And so unborn babies hiccup for about half an hour every day until they are born. However, in adults the hiccup seems to have no purpose, except to be really annoying.

In 1988, an extreme case of hiccuping entered the life of Dr Francis M. Fesmire while he was working in the Emergency Department of the University Hospital in Jacksonville, Florida. Initially, he was stumped as to what treatment he could offer his hiccuping patient. But in the time-honoured Medical Tradition of "See one, do one, teach one", he remembered reading something relevant, tried it out, and wrote it up as a cure for hiccups. (Mind you, not everybody will resort to "taking" Dr Fesmire's "medicine" – read on, and you'll soon know why.)

There are well over 100 different causes of hiccups.

You can divide these causes into six major groups. They are:

- disorders of the Peripheral Nervous System
- disorders of the Central Nervous System
- metabolic or drug-mediated causes
- infectious causes
- psychogenic causes and
- unknown causes.

For example, epidemic hiccups were recorded during the influenza-encephalitis outbreaks in the 1920s. Alcohol, various foods (including carrots) and gastric distention can also cause hiccups.

HICCUPS 101

There is a wise old saying in medicine: "If there is one single cure for something it probably works, but if there are many cures they hardly ever work."

Hiccups are an excellent example of this. There are almost as many "cures" as there are possible causes. Cures include drinking water from the wrong side of a cup, breathing in and out of a paper bag sprinkled with vinegar, and chewing frozen raspberries. If you are lucky enough to find something that works for you, be happy – but don't assume that your cure will work for everybody else.

So what is hiccuping? Well, it's more than a simple spasmodic contraction of the diaphragm.

The short explanation is that a hiccup happens when your diaphragm muscle contracts abruptly. So you breathe in suddenly, and a little bit later your vocal cords snap shut. This interrupts the airflow into your lungs, giving the characteristic "hic" sound of the hiccup.

> There is a wise old saying in medicine: "If there is one single cure for something it probably works, but if there are many cures they hardly ever work."

The longer explanation is that the hiccup is a complex, patterned and coordinated motor act of several groups of muscles. It involves your mouth, diaphragm, lungs and larynx.

The very first muscle activity in a hiccup is that both the roof of the mouth and the back of the tongue begin to lift.

Then the diaphragm gets involved. The diaphragm is a curved muscle in the shape of a half-dome. At the front of the chest it connects to the bottom of your ribs and then it curves back and downwards to join onto your lower back. It separates your lungs, which are above it, from your gut, which is below it. When it contracts, it pulls your lungs downwards, creating suction so air enters your lungs.

Finally, the larynx (or voice box) in your throat joins the hiccuping party. In your larynx are the vocal cords which can be fully open, fully shut and anywhere in between. (The "glottis" is the combination of the vocal cords and the space between them.)

Millions of years ago in evolution, your larynx "captured" control of your flowing air, so that you could make speech. As the air leaves your lungs it passes through the larynx and then through your mouth. Both your larynx and your mouth do very complicated adjustments to the outgoing air to make speech.

The hiccup starts in what seems to be a Hiccup Reflex Control Centre (HRCC) in the spinal cord, between the third and fifth cervical segments. The phrenic (related to diaphragm) and vagus nerves come into the HRCC, as does the sympathetic chain (T6–T12). Leaving the

HRCC is another branch of the phrenic nerve. Electrical impulses travel down this branch to the diaphragm and tell it to contract, making you breathe in. Other electrical impulses go to muscles that are involved in breathing out and switch them off. About 35 milliseconds (thousandths of a second) after the air starts flowing in, an electrical signal is sent to the vocal cords, making them snap shut. (This usually coincides with the heart contracting.) And, *voilà*, a hiccup is felt and heard.

Longest Hiccuping

Charlie Osborne began hiccuping in 1922, while he was slaughtering a pig. The hiccups finally stopped 68 years later on 5 June 1990, after his youngest daughter passionately embraced three days of "fierce praying". He died 11 months later.

TREATING HICCUPS

As the Vagus Nerve is implicated in causing hiccups, perhaps fooling around with the Vagus Nerve could help you treat them.

The Latin word "vagus" means "wandering". ("Vagus" also gives us the English words "vagrant", "vague" and "vagabond".) The vagus is actually one of the 12 cranial nerves but, true to its name, it wanders around the body. It travels into the chest cavity, where it is involved with the lungs and heart, and also deep into the gut cavity, from the mouth to the anus. The Vagus Nerve coordinates swallowing and breathing, and it even runs the vocal cords.

One moderately reliable treatment for hiccups involves deliberately overstimulating the Vagus Nerve, which will block other signals to the vocal cords.

Why Hiccup?

In today's humans, the act of hiccuping seems
to have no useful function. So why do we hiccup?

Dr Christian Straus at Pitié-Salpêtrière Hospital in Paris
reckons that it began, and was useful, 370 million years
ago when our ancestors left the oceans for the land. Those
primitive air-breathers still had gills, and pushed water across
these gills by squeezing their "cheeks" – this closed the
glottis so that water would not enter the lungs.

Perhaps when unborn babies are hiccuping in the uterus,
they are practising for sucking milk once they are
born – when they close their glottis to keep the
milk out of their lungs.

So perhaps hiccuping as adults is the leftover
penalty for being able to suck milk and breathe
at the same time when we were babies?

HICCUPING MAN

An otherwise healthy, muscular 27-year-old man turned up in Dr Fesmire's Emergency Department with a 72-hour history of constant hiccuping at the rate of 30 per minute. His past health was unremarkable, although he did have a lifelong history of frequent

One moderately
reliable treatment
for hiccups
involves deliberately
overstimulating
the Vagus Nerve.

episodes of hiccuping. However, none of them had ever lasted more than two hours, and all of them had resolved spontaneously. This episode had been running for 72 hours and he was exhausted.

To treat him, Dr Fesmire started trialling a repertoire of manoeuvres well known to stimulate the Vagus Nerve. Setting off the Gag Reflex by touching a tongue depressor to the back of the throat did nothing for his patient's hiccups. Neither did pulling on his tongue. But the Valsalva Manoeuvre (where you try to blow air out from your lungs while you block your mouth and nose) did slow the hiccups down from 30 per minute to 15 per minute – but only while actually performing the manoeuvre. As soon as the patient finished the Valsalva Manoeuvre, the hiccup rate went back to 30 per minute. The same tantalising partial response – a brief slowing, followed by return to the original hiccup rate – happened for both massaging the carotid sinus in the side of the neck, and for physically compressing the eyeball with the fingers.

But then Dr Fesmire remembered reading a paper the previous year with the title of "Termination of Paroxysmal Supraventricular Tachycardia by Digital Rectal Massage". (Paroxysmal Supraventricular Tachycardia involves the heart accelerating up to 200 beats per minute, and it can be very uncomfortable – shortness of breath, chest pain, dizziness, loss of consciousness, etc.) He theorised that if this unconventional treatment worked for the heart it might work for hiccups, because both are influenced by the Vagus Nerve. As Dr Fesmire wrote, "Digital rectal massage was then attempted using a slow circumferential motion. The frequency of hiccups immediately began to slow, with a termination of all hiccups within 30 seconds. There was no recurrence of hiccups during the next 30 minutes and the patient was discharged ..."

Hiccup in Other Languages

Most of the words for "hiccup" in other languages sound like an actual hiccup. The Spanish word is "hipo", French "hoquet", Russian "ikota", Dutch "hik", Kurdish "hirik", and in Arabic it's "hakka".

Why did this work?

According to Dr Fesmire, "the rectum is supplied with an abundance of sympathetic and parasympathetic nerves, and the digital rectal massage would lead to increased vagal tone and potential termination of hiccups". Sensation from the rectum travels through these parasympathetic nerves – and they are very sensitive to pressure.

After reading this report, Dr Odeh from the Bnai Zion Medical Center in Haifa, Israel, found that this technique worked to fix the hiccups of a 60-year-old man with acute pancreatitis. Dr Odeh then went on to use this technique successfully in five other patients with intractable hiccups.

In 2006, Dr Fesmire was honoured with the Ig Nobel Prize in Medicine at a ceremony at Harvard University in Boston, Massachusetts. He later told *New Scientist* magazine of a treatment likely to be more popular with hiccuping patients: "An orgasm results in incredible stimulation of the vagus nerve. From now on, I will be recommending sex – culminating with orgasm – as the cure-all for intractable hiccups."

But first, check if your orgasm is covered by your health insurance . . .

Fatal Hiccups

Hiccuping can, via a circuituous and indirect pathway, lead to death (but only very rarely).

In one case, "a 23-year old man fatally struck his friend in the chest with a closed fist, as a mutually agreed on remedy for hiccups". By an accident of terrible timing, this blow was able to set the friend's heart into Ventricular Fibrillation, from which he died. This is called "Commotio Cordis", or "Shaking Heart".

The *Queensland Times* reported a similar case on 5 July 2000.

A different pathway to death via hiccups happened in the port city of Barranquilla in Colombia on 22 January 2006. David Galvan, 21, and his uncle, Rafael Vargas, were drinking with neighbours when David began hiccuping. Rafael pulled out his revolver to scare David's hiccups away – but accidentally shot him dead. Greatly distressed, Rafael then committed suicide by turning the gun on himself.

Cures for Hiccups

Ancient healers such as Celsus, Galen and Plato
tried to treat hiccups – with varying success.

Today's treatments fall into three main categories.

The first category is physical manoeuvres, such as swallowing
a teaspoon of granulated sugar, stimulating the back of the
throat, performing the Valsalva Manoeuvre or Carotid Sinus
Massage, generating digital eyeball pressure, fright, etc.
The second category is drugs such as metoclopramide,
diazepam, haloperidol, etc, and the third is miscellaneous
treatments such as hypnosis, psychotherapy or high
carbon dioxide levels.

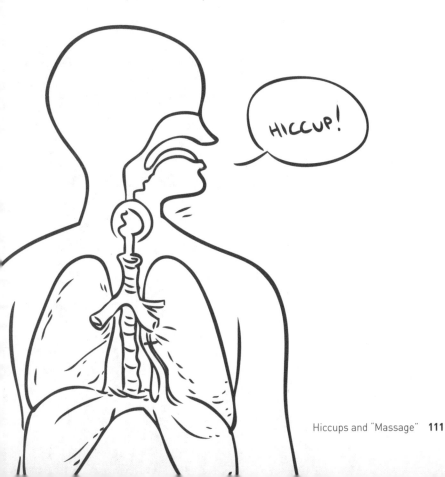

18

JUDGES' DECISIONS: MEALS AND APPEALS

One thing that I learned as a medical doctor is that making decisions can be very hard. Judges do it tough, because they have to make hard decisions all the time. You would expect that these decisions would be fair and unbiased, and not influenced by extraneous factors. And that's why I was surprised after I read a scientific paper actually called "Extraneous Factors in Judicial Decisions".

This paper claimed that a judge's decisions depended mostly on how much time had passed since the judge last took a break. This blew me away.

BIASED OR FAIR

> "The life of the law has not been logic: it has been experience." The plain English version is: "Justice is what the judge ate for breakfast."

In western society our courts supposedly embody the principle of Legal Formalism. This implies that judges should apply formal legal reasoning to the specific circumstances of a case. Doing this in a rational, mechanical and deliberative manner should lead to fair and equitable justice for all.

However, the Legal Realist movement disagrees. United States Supreme Court Justice Oliver Wendell Holmes contributed to this highly influential 20th-century point of view when he said: "The life of the law has not been logic: it has been experience." Psychological, social and political factors can also influence the judge, claims Legal Realism. The plain English version of Legal Realism is: "Justice is what the judge ate for breakfast".

DECISIONS ARE HARD

Making decisions is hard.

Military, medical and legal people have to make far-reaching decisions. It is well known that even if you do possess enough information, having to repeatedly make decisions or judgements will deplete your brain's executive function and mental resources.

It's not just at work that you're expected to make repeated decisions. You might have had to buy a new white good recently – and found yourself simply worn down by the process. A white good is usually known as a "Grudge Purchase".

A Grudge Purchase is not something that you really want, such as wonderful and exciting clothes or a new and shiny and incredibly useful electronic gadget. No, it's something that you need rather than love – such as a washing machine or a lawnmower. So first you check out all the relevant consumer reports, and then go to buy a specific model. But it turns out that the specific one you want is not available, so you choose something similar. Unfortunately, your second choice won't be available for a few weeks, and same for your third choice. Eventually you give into the inevitable, avoid making another decision, and just take whatever washing machine the shop has in stock on that day.

Parole, Life Sentences and Juveniles

The USA, the "Land of the Free", has the highest rate of incarceration in the world. In 2009, 2.3 million people were in American prisons. This figure has grown some 600 per cent since 1972. It's the result of "three decades of 'tough on crime' policies that have made little impact on crime but have had profound consequences for American society". These policies emphasise drug enforcement, mandatory sentences, a vastly expanded use of imprisonment and widespread cutbacks in parole.

Parole has always served as an incentive for prisoners to improve their behaviour. It's also a measure of their suitability to be returned to society.

But in the USA, 30 per cent of the prison population who are serving a life sentence have no possibility of parole. According to the *New York Times*, "from 1992 to 2008, the number in prison for life without parole tripled from 12,453 to 41,095, even though violent crime declined sharply all over the country during that period".

This is especially tough for juveniles, who often go along with their friends simply because they want to fit in.

In the case of juveniles who are serving a life sentence without parole in the USA, about 60 per cent were first-time offenders. About one quarter of all juveniles serving a life sentence without parole were not the primary assailant. In many cases, they were present at the crime scene, but only minimally involved. However, they were automatically given a life sentence without the possibility of parole, because of the wording of the local US state law.

In 2010, the US Supreme Court ruled that sentencing a juvenile to life imprisonment without parole (when the crime is not homicide) is a cruel and unusual punishment. Justice Anthony Kennedy said that a life sentence without parole shares "some characteristics with death sentences that are shared by no other sentences . . . (in altering) . . . the offender's life by a forfeiture that is irrecoverable".

THE PAROLE STUDY

This study specifically looked only at prisoners who were already in jail and who were applying for parole. Parole is early release under supervision. The sample size was large enough to do good statistics.

There were eight Jewish–Israeli judges (six male, two female) with an average experience of 22 years. Over a 10-month period, they made 1112 parole rulings over 50 sitting days. Between them in that period, these judges made about 40 per cent of all the parole rulings in the state of Israel. The prisoners were about two-thirds Jewish–Israeli males, one-third Arab–Israeli males. The crimes that the prisoners had committed included embezzlement, assault, theft, murder and rape.

There are many factors that you might reasonably expect to influence a judge's decision. These could include the number of previous incarcerations of the prisoner, the gravity of the crime committed, the time already served in prison, the availability of a rehabilitation program, and the age, gender, religion and nationality of the prisoner.

Most of these factors had no influence on the judges' decisions. The number of previous incarcerations and the availability of rehabilitation programs were the only two factors that had even a little influence.

TIME IS ON MY SIDE . . .

But, overwhelmingly and incredibly, the most powerful and significant factor affecting their decisions was simply the length of time that the judge had spent judging since their last meal break. I find this absolutely astonishing.

The judges had two meal breaks each day. So they had three separate sessions each day, separated by a 20-minute morning snack and a 60-minute lunch.

First thing each morning, the judges would grant parole to about 65 per cent of the prisoners applying for it. As the morning wore on, this would zigzag relentlessly downward to 0 per cent. Then they would have their morning snack.

After the morning break, the judges would again grant parole to 65 per cent of the applicants. This would bumpily drop down to around 10 per cent. Next it was lunchtime.

After lunch, the judges would again grant parole to 65 per cent of the applicants. This would plummet quickly to around 10 per cent, bounce around this level as the afternoon wore on, and then drop to 0 per cent. Finally, it was time to go home.

TIMING IS EVERYTHING

Surprisingly, nobody involved in the parole system had any idea that this was happening. "Nobody" included the prisoners' attorneys and those on the parole board – criminologists, social workers and judges.

The authors summarised their paper with, "We find that the percentage of favorable rulings drops gradually from around 65 per cent to nearly zero within each decision session and returns abruptly to around 65 per cent after a break."

They also said, "our findings suggest that judicial rulings can be swayed by extraneous variables that should have no bearing on legal decisions".

Perhaps the word "suggest" is a little mild. Maybe the judges were running out of willpower as the time increased since their last break and so were taking the "easy" option of not changing anything (by refusing parole).

And perhaps, if you or a relative or friend should end up in court before a judge, it might be worthwhile getting friendly with the court official in charge of schedules – or bringing cupcakes for the judge.

Why is it so?

What does this finding mean for people who have to make a whole bunch of major and separate decisions, one after the other, such as medical doctors and those working in financial markets? Should they eat something at well-defined intervals during the day, or do they just need a break?

We don't know what caused the granting of parole levels to jump from around 0 per cent to 65 per cent immediately after the break. Was it the act of having a break, or was it the

ingestion of food leading to higher blood glucose levels?
(See "Marshmallows, Money and Munchies", page 1.)
Did they do some exercise, drink lots of caffeine, or did
they gaze longingly onto an open grassy field? Hopefully
follow-up studies will give us the answer.

Hard to be Easy, Easy to be Hard

In the study involving our eight judges, it was easier for them
to reject a request for parole than to allow it. Rejection of
parole simply meant not changing anything and leaving the
prisoner in jail – in other words, keeping the status quo.

This rejection of parole happened far more frequently when
the judges were tired or mentally depleted – such as after a
long session of sequential decision-making.

We know that it was harder to grant parole because of two
facts. First, it took the judges longer to write a favourable
ruling that granted parole (7.4 minutes) than an unfavourable
ruling (5.2 minutes). Second, there were more words in the
written verdicts of favourable rulings (89.6) than
unfavourable rulings (65.5).

Society and Prisons

We humans are probably the only animal that has so-called Third-Party Punishment – "I punish you because you harmed him".

First-Party Punishment protects your immediate family, and your DNA. In this case, you retaliate against people who harmed you or your immediate family.

Second-Party Punishment works only in relatively small groups, where unrelated people protect each other for the common good. The fancy names are Reciprocal Altruism or Direct Reciprocity. The non-fancy versions are "you scratch my back and I'll scratch yours" (think of Mama Morton in the movie *Chicago*) and its obverse, "I punish you because you harmed us". Chimpanzees carry out Second-Party Punishment, but not Third-Party Punishment.

Third-Party Punishment helps create a stable society that will eventually improve the overall survival of your family's DNA. It starts very young. Even children as young as three engage in Third-Party Punishment. It relies upon impartial, state-empowered enforcers to punish those who break the rules of the society. In this situation, people are willing to punish those who steal or violate other social rules, even though they were not directly harmed and will not directly benefit from the punishment.

19

MICROWAVE OVEN HERTZ YOUR WiFi

One question that comes up every few months on my ABC and BBC radio shows is why the WiFi/interweb system stops working whenever a microwave oven is used nearby. It's because the microwave oven and the WiFi use very similar frequencies – and the much more powerful microwave oven overcomes the much weaker WiFi.

WiFi 101

WiFi is just a way to deliver access to the interweb without cables. It was "invented" in its earliest version back in 1991. I first used it in 2000 and was amazed. (I still am.)

The "WiFi box" or "Access Point" is just a gutless little radio transmitter – by "gutless" I mean a maximum power of only one-tenth of a watt. It broadcasts a radio signal with a frequency of around 2.45 gigahertz. The radio signal carries the interweb information to your computer, which strips away the 2.45 gigahertz carrier signal – and suddenly, without any wires, you are surfing the web.

> The "WiFi box" or "Access Point" is just a gutless little radio transmitter.

But it's a little bit more complicated than this.

First, there are about a dozen different frequencies all next to each other (and, surprisingly, overlapping each other) around that nominal 2.45 gigahertz. This means that a bunch of computers and WiFi boxes can all surf the web at the same time without interfering with each other too much.

Second, some WiFi boxes (and computers) can talk to each other on another set of frequencies up around 5 gigahertz. These frequencies don't experience interference from microwave ovens. On the other hand, their range is shorter.

MICROWAVE VERSUS WiFi

To heat up food, microwave ovens pump out about 1000 watts – about 10,000 times more than a WiFi Access Point.

You need only the tiniest amount of leakage to cause interference. This microscopic leakage is not dangerous to humans, but can really muck up your web access – so that pages either load very slowly, or totally freeze.

A technical writer, Jim Geier, tested a WiFi Access Point at varying distances from a microwave oven while it was working.

At 30 centimetres from a working microwave oven, the data rate dropped by 90 per cent. It was down by 75 per cent at a distance of 2–3 metres, and by 60 per cent even at 6 metres.

It was worse again if there were several users on the network.

So what can you do?

THE FIX

First, keep the WiFi unit well clear of your microwave oven. Ten metres would seem to be the minimum.

Second, the problem is a lot worse with the higher-numbered WiFi channels (such as 8–11), as these channels are closer to the frequency that most microwave ovens run on. So try changing to the lower numbered channels. (You'll have to log in to the WiFi Access Point with software – good luck.) Check the label on the back of the microwave oven to find out exactly which frequency to avoid.

Or if these measures don't work, try weaning yourself from the interweb when you are preparing dinner.

WiFi and Baby Monitors

Many other devices also use the same
2.45 gigahertz frequency band
that WiFi networks do.

These devices include cordless phones, remote doorbell
ringers, Bluetooth devices, internal movement sensors
in some car alarm systems, video devices that send
a video signal from one place (surveillance camera
on roof, computer, etc) to another,
and yes, baby monitors.

All of these devices can interfere with your
WiFi network – and sometimes your
WiFi network can interfere with them.

20

LUNAR LUNACY: A WANING THEORY

The modern plethora of werewolf books, TV series and movies are idealogically consistent with the 1941 Hollywood classic film *The Wolf Man*. Yep, in the movies, if you are susceptible, the Full Moon will still turn you into a lunatic werewolf.

In fact, the delightfully antiquated word "lunacy" comes from Luna, who was the Roman Goddess of the Moon. One definition of "lunacy" is "intermittent insanity once believed to be related to phases of the moon".

The belief that Lunacy is linked to the Moon goes back a long way. The Roman scientist and military commander Pliny the Elder said that the Full Moon causes a very heavy nocturnal dew, so it must also make the brain "unnaturally moist".

Actually, the Full Moon has virtually no effect on our minds.

Lunar Lunacy in Literature

In *Othello*, Shakespeare wrote,
It is the very error of the moon,
She comes more near the earth than she was wont,
And makes men mad.

MOON MAKES MADNESS?

US surveys in 1985, 1990 and 1995 found that about 40 per cent of the general population, and (incredibly) 80 per cent of mental health professionals, believe that the phase of the Moon affects human behaviour. They believe this even though 99.9+ per cent of the evidence says that the Full Moon has no effect on human behaviour.

The Moon takes just under a month to run from New Moon (darkest), to half full (First Quarter), to Full Moon (brightest), to half full (Third Quarter) and back to New Moon again.

But it's the Full Moon that is believed to be linked to a huge list of human misery, including accidents, alcoholism, anxiety, assaults, calls to crisis telephone numbers, casino activity, depression, domestic violence, drug overdoses and, of course, Emergency Room visits. If that's not enough, it's also supposedly responsible for human-made disasters (plane or train crashes, etc), illegal drug use, kidnappings, murders, natural disasters, prison violence, psychiatric disturbances, psychiatric patient admissions, self-harm, shooting incidents, stabbings, suicides, the amount of food we eat and so on.

Over the last half-century, thousands of studies have looked at the Moon's effect upon all these behaviours. An occasional study will show a correlation with the fullness of the Moon – but then, more thorough follow-up studies in the scientific literature disprove it.

But this is very different from what will appear in your local newspaper, or on your TV. After all, journalists have a story to manufacture and a deadline to meet – so they won't let the facts get in the way.

Science of Lunar Lunacy

Studies that look at large numbers and long periods of time almost always show no Lunar Lunacy effect. For example, a US study in 1983 surveyed 361,580 calls for police assistance – and there was no link to the phase of the Moon. A study of 54,547 trauma admissions to the Emergency Rooms of three hospitals in Iran showed a complete lack of Lunar Lunacy. A US study looked at 185,887 attempted suicides over a seven-year period – again, no Lunar Lunacy.

Yet a tiny number of studies do show Lunar Lunacy. Sometimes this is just a normal statistical bump. But sometimes the authors made a silly mistake. For example, one study supposedly showed that car accidents were more common during the Full Moon. Closer analysis showed that in that specific location on the weekends, more people were on the road at night. Also, over the time period chosen by the authors, Full Moons were more common on the weekends. When these factors were accounted for, the Lunar Lunacy Effect vanished.

Unfortunately, it's the statistically odd studies that get the headlines in the popular press.

REAL-LIFE LUNAR LUNACY

Before artificial lighting, people stayed up later during the Full Moon. After all, when the Full Moon is hanging in the sky, the night is 250 times brighter than if there's no moonlight at all. So, even today, in so-called "primitive" societies that don't have artificial lighting at night, the Full Moon is the occasion for a party, revelry and a general good time. If there are more people up and about, then obviously mishaps (falling over, etc) will be more frequent. Having more people around does mean more human activity – both good and bad.

But in our modern, technological, artificially lit society, the Moon does not make people go mad, nor does it increase numbers at hospital emergency rooms, nor does it increase suicide rates.

BIOLOGICAL TIDES THEORY

Even though the Lunar Lunacy Effect does not exist, there is a theory to explain it!

It's the Biological Tides Theory. This crazy "theory" says that the Moon has a huge effect on the tides, which are made of water. The human body is mostly made of water (the "theory" continues), therefore the Full Moon must have an effect on us.

This "theory" is wrong in a few ways.

First, the "Moontides thingie" happens because the oceans are made of liquid. Tides would still happen if the liquid was freezing liquid hydrogen, room-temperature mercury or boiling liquid iron. The liquid doesn't have to be water.

Second, tides happen only over large expanses, not within the small dimensions of a human body.

Third, the ocean tides happen regardless of the phase of the Moon – there are two of them every single day. We still have high and low tides every day, regardless of whether the Moon is full, new or half full. The Moon still has a gravitational effect even if the Sun doesn't fully light it up for us.

Mosquito Lunacy Effect

The late astronomer George Abell was quoted as saying that a mosquito sitting on our arm exerts a more powerful gravitational pull on us than the moon does. He was right.

But there have never been any reports of a Mosquito Lunacy Effect.

SELECTIVE RECALL THEORY

A better theory to explain Lunar Lunacy is Selective Recall. Maybe it's a busy night at work, and you look out the window for a brief respite and see the very conspicuous Full Moon. You put two and two together to make five, and wrongly assume that the Full Moon made your night busy – just because it stood out, and you selectively took notice of it.

This belief that the Full Moon massively affects human behaviour is a cultural fossil. It's a memory of the way that we would party on the night of a Full Moon, way back when we had no artificial light. But in our modern society, it's just moonshine. (That's "moonshine" in the sense of "nonsense you speak on a Full Moon".)

Real Moon Effects

When the Moon is full, all of its surface that we can see is illuminated by the Sun. The Moon re-emits heat (or infra-red rays) at us. So, every Full Moon, our lower atmosphere gets hotter by 0.02 degrees Celsius than at a New Moon.

This might explain why the night of the Full Moon is on average cloudier than other nights.

There is also a weak link to earthquakes. When the Sun, Earth and Moon align, this makes our planet flex by about 23 centimetres (out of a total diameter of about 12,700 kilometres). This sets up stresses in the crust that are about 1000 times weaker than the normal stresses found in the crust. If there's an earthquake that's about to pop, this microscopic extra stress might be the final straw needed to set off the earthquake.

And, finally, pregnancy. A Synodic Month is the 29.53 days between one Full Moon and the next. The length of the average pregnancy (266 days) is almost the same as 9 Synodic Months (265.77 days). So, if you love another person Very Much in a Special Way at a certain phase of the Moon, then on average, that will be the Moon's phase for the birth.

21

SMOKING AND WEIGHT LOSS: THE PLAIN TRUTH PACKAGED

We know that, on average, smokers are thinner than non-smokers. We also know that when smokers quit, many of them gain weight.

So how can smoking cause weight loss?

Is it behavioural? Do people smoke instead of eat? After all, when I used to smoke cigarettes, I would spend less time eating to give me more time for smoking.

Or is it pharmacological? Does nicotine vastly increase your overall metabolic rate, so you burn up huge amounts of energy? Or does nicotine react directly on nerves somewhere in your brain to reduce your appetite?

The answer seems to be mostly appetite reduction, with a bit of increased metabolic rate.

NICOTINE 101

Nicotine is an alkaloid chemical found naturally in the nightshade family of plants. These plants make nicotine in their roots and then send it up to their leaves. In the leaves, nicotine acts as a very effective poison to stop insects that live on the plants from eating them. So tobacco plants, in which nicotine makes up about 2 per cent of the dry weight, tend to be pretty resistant to most insects.

Naming Nicotine

Nicotine gets its name from Jean Nicot de Villemain, who was a French ambassador. He sent some tobacco and tobacco seeds to Paris back in 1560.

It took until 1828 to isolate nicotine from the tobacco plant. It's a fairly simple chemical with 14 hydrogen atoms, ten carbon atoms and two nitrogen atoms. Back then, the technology to analyse chemicals was in its infancy. Nicotine's formula was worked out in 1843, but the 3-D model of the atoms wasn't worked out until 1893. Finally, in 1904, it was first made artificially in a laboratory.

There are many ways to get nicotine into your brain. You can smoke it, snort it, chew it, rest it against your gums or absorb it through your skin from a transdermal patch. If you smoke it, it quickly enters your bloodstream and within 10–20 seconds it has crossed the blood–brain barrier and entered nerve cells in your brain.

Nicotine has very many different ways of affecting your body.

When it's present in low levels in your blood, it can act on the adrenal glands to release more adrenaline into your body. This can simultaneously stimulate you and make you more anxious.

There are many ways to get nicotine into your brain. You can smoke it, snort it, chew it, rest it against your gums or absorb it through your skin from a transdermal patch.

But at high blood levels, nicotine can be a mild sedative, thanks to its activity on your native serotonin and endorphin chemicals.

And in very high blood levels, say 1 milligram per kilogram of body weight, it can kill you.

But even at more average blood levels, nicotine can alter levels of natural chemicals in your body such as dopamine, vasopressin, arginine, acetylcholine and so on. It can constrict your blood vessels and force your body to dump cholesterol into the bloodstream.

HOW NICOTINE MAKES YOU SKINNY

So there are many complicated ways in which nicotine can affect your body. So it's understandable that it took us until mid-2011 to work out how nicotine keeps you skinny. A team of American and Canadian scientists performed a rather neat and sophisticated study. They had to use a combination of pharmacological, electrophysiological, molecular genetic, and feeding studies to understand exactly what was happening.

They worked out that nicotine firstly stimulates some specific cells in your brain. These then stimulate a different, but particular, set of cells elsewhere in your brain, which finally have a typical effect on your behaviour. This results in you eating up to 50 per cent less. This in turn reduces your body fat mass by about 15–20 per cent. There also seems to be a small increase in metabolic rate, which burns up some extra energy.

TECHNICAL STUFF

A part of the brain called the arcuate nucleus (ARC) is heavily involved in regulating feeding behaviour. Nicotine seems to activate pro-opiomelanocortin (POMC) neurons in the ARC. Once activated, these cells then increased activity in the hypothalamus, which then decreased eating.

More specifically, the scientists discovered some of the intermediate chemicals from this process in your brain linked to the weight loss associated with smoking. In the future, we may find ways to stop the well-known increase in weight that happens in many people who give up smoking. In fact, we might even find some drugs that could help control obesity.

But don't even think of using smoking as a way of losing weight. Smoking is so bad for your health in so many different ways that it would take me an hour just to give you a summary. After all, half the people who smoke cigarettes long term will die from a smoking-related disease.

Big Tobacco

Nicotine addiction is one of the hardest addictions to break. To make it even harder, Big Tobacco in the USA increased the nicotine content of cigarettes by 1.6 per cent per year between 1998 and 2005. This trend happened in all major cigarette categories – mentholated versus non-mentholated, and full flavour versus light, medium/mild or ultralight.

Don't ever trust the tobacco industry, because, as Dr Stanton Glantz from the University of California said when he compared the leaders of the tobacco industry with cockroaches, "They love the dark, and they spread disease".

Big Tobacco in the USA increased the nicotine content of cigarettes by 1.6 per cent per year between 1998 and 2005.

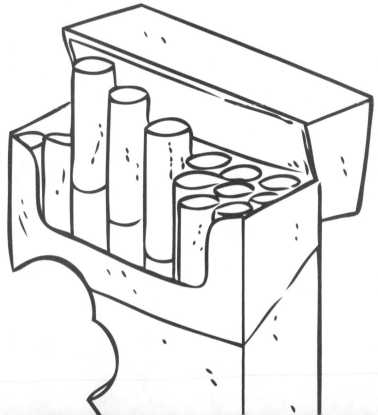

22

THE ANNE HATHAWAY EFFECT: MOVIE STAR AND MONEY MARKETS

We normally think of Hollywood movies, and those who act in them, as purely providing entertainment. But Anne Hathaway is different – whenever she hits the headlines, she also hits the share price of one of the biggest companies in the world. (Please note the following before acting on any financial advice offered in this story – I am not a registered financial advisor.)

It's because of two separate factors. The first one is her surname, Hathaway. The second one is the Stock Market, and more specifically, High Frequency Trading, and Analysis of Market Sentiment.

HIGH FREQUENCY TRADING

Let's deal with High Frequency Trading first.

Back in the old days, trading stocks or shares was straightforward. The potential buyers and sellers would meet on the floor of the Stock Exchange and bargain with each other until they made a deal. But in 1998, the US Securities and Exchange Commission changed the rules so that electronic exchanges could compete with traditional exchanges such as the New York Stock Exchange. All you needed was a computer.

Today, powerful computers can scan dozens of stock markets simultaneously, and can find trends in buying or selling literally more quickly than a human can blink. They then execute millions of "buy" or "sell" orders in less than a second. Typically, High Frequency Traders would hold each stock for a very short time (seconds to hours), and might buy and sell that stock many times each day. They make only a tiny profit on each transaction, but they carry out tens of thousands of transactions each day.

> A handful of High Frequency Traders count for more than 60 per cent of the 7 billion shares traded each day on the US stock markets.

The detractors of High Frequency Trading say that the speed and volume of the trades can distort the prices of stocks and make financial markets less stable. They also say that the traders with the most powerful supercomputers exclude the ordinary investors, because they can enact the best deals more quickly. High Frequency Traders counter by saying that they have cut trading fees, and have brought more competition to the markets.

Today, a handful of High Frequency Traders account for more than 60 per cent of the 7 billion shares traded each day on the American stock markets. It's very profitable for them.

0.03 Seconds = Billions

In 2010 and 2011 in the USA, High Frequency Traders generated about US$12.9 billion in profits.

Here's an example of how they did it. On 15 July 2009, for various reasons, shares in the semiconductor company Broadcom were likely to rise. The people who knew this were the slower traders, who used the stock market in the traditional way. If they bought one huge block of shares, this would attract attention and probably increase the price of Broadcom shares. So, to cover their tracks, they broke up the order into many dozens of smaller batches.

In the Old Days, this would have worked.

But the American Nasdaq Stock Exchange has a deal where, thanks to a loophole in the regulations, traders can get an advance peek at who is buying or selling – for a suitable fee, of course. The advance notice they pay for is not very much – only 0.03 seconds. But that was enough for the computer programs (or algorithms) of the High Frequency Traders to pick up the growing hunger for Broadcom shares. In less than 0.5 seconds, these algorithms had already begun buying shares at $26.20, and then reselling them to the slower human traders.

But the algorithms had another trick up their sleeves. They began issuing "buy" and "sell" orders for Broadcom at a whole range of prices, and then cancelled these orders almost immediately. Very quickly, the algorithms worked out that some investors were prepared to pay up to a maximum of $26.40. And so the algorithms, without any

human involvement, started selling tens of
thousands of shares at $26.39.

In this case, the slower-moving human investors (who bought
at $26.39 instead of $26.20) paid about $7800 more than if
they too had High Frequency Trading algorithms. $7800
is not bad for a few minutes' work, especially if
you can do it thousands of times every day.

ANALYSIS OF MARKET SENTIMENT

So now let's look at the second factor, Analysis of Market Sentiment.

A very important way for investors to work out whether to buy or sell shares in any particular company is by reading the Chief Executive Officer's speeches, the company's annual reports, newspaper articles and so on. But it would take a person a long time to do this. So computer programs (or algorithms) have been written to trawl through all the relevant literature, looking for positive or negative feelings about any specific stock. They don't try to understand the article – they are merely hunting for positive or negative sentiments that relate to the companies they're interested in.

So how does this all link back to Anne Hathaway?

It turns out that one of the wealthiest men on the planet, Warren Buffet, runs a very wealthy company called Berkshire Hathaway. Every year, it's usually in the top ten public companies in the world. Over the last 44 years, the company has grown by over 20 per cent each year, and currently has revenue of over $140 billion per year, and employs over a quarter of a million people. A single share in Berkshire Hathaway costs over US$120,000.

And every time Anne Hathaway makes headlines, the stock price of Warren Buffett's Berkshire Hathaway climbs. It doesn't matter whether it's the opening of her latest movie or if she co-hosts the Oscars – Berkshire Hathaway climbs a few per cent.

Class Warfare

In 2006, Warren Buffet said that it was very unfair that he paid only 19 per cent income tax, while his much poorer employees paid 33 per cent. He said, "There's class warfare, all right, but it's my class, the rich class, that's making war, and we're winning."

CONFUSION

Anne Hathaway has a good reputation, and her movies, such as *Brokeback Mountain*, *The Devil Wears Prada* and *Alice in Wonderland*, usually do well. So media reports involving the name "Hathaway" are usually positive. The Analysis of Market Sentiment computer algorithms link "Anne Hathaway" with "Berkshire Hathaway". The Analysis of Market Sentiment algorithm feeds data into the High Frequency Trading algorithm, which then buys some Berkshire Hathaway stock – even at over US$120,000 per share.

On one hand, High Frequency Traders make their wealth by methods that add no value to our society – only to their own pockets. On the other hand, Warren Buffett has always looked for companies that have strong fundamentals and are in business for the long term, not the quick profit. He has pledged to give away 99 per cent of his wealth to charity. His house is modest and he lives in Middle America in Omaha, Nebraska, and is fondly called "The Oracle of Omaha".

How funny that High Frequency Traders can't tell the difference between The Oracle and The Actress.

Hathaway Effect

This list of Anne Hathaway movies and subsequent movement in "BRK.A" (Berkshire Hathaway stock) is based on the Dan Mirvish article in the *Huffington Post.*

3 October 2008	*Rachel Getting Married* opens: BRK.A up 0.44 per cent.
5 January 2009	*Bride Wars* opens: BRK.A up 2.61 per cent.
8 February 2010	*Valentine's Day* opens: BRK.A up 1.01 per cent.
5 March 2010	*Alice in Wonderland* opens: BRK.A up 0.74 per cent.
24 November 2010	*Love and Other Drugs* opens: BRK.A up 1.62 per cent.
29 November 2010	Anne announced as co-host of the Oscars: BRK.A up 0.25 per cent.

On the Friday before the Oscars, Berkshire shares rose a whopping 2.02 per cent. And on the Monday just after the Oscars, they rose again, this time 2.94 per cent.

23

MILK
MAGNIFIES
MUSCLE

In the human body, muscle is a remarkable tissue with some magical properties. For one thing, it turns chemical energy into motion and force. For another, it's very plastic, or changeable. If you do nothing, your muscles shrink. But if you work your muscles hard, they change shape and get bigger and stronger.

People have spent a lot of time and energy working out the best way to Pump Iron in a gym, so they can bulk up as quickly as possible. But how can you bulk up even more immediately after exercise – when you've stopped Pumping Iron – and apparently are doing nothing at all?

It turns out that what you eat or drink in that important (but often neglected) period after you finish Pumping Iron is crucial to building up more muscle.

MUSCLE 101

Overall, there are three main types of muscle in the body.

First, there's cardiac muscle, in your heart. It's in there for the long term. Basically, it starts working before you're born, and when it stops, you die.

Second, there's smooth muscle. It's found inside your gut, and makes up the walls of the 10 metres of tube that run from your top to your bottom. It pushes your food from your mouth to your anus by cleverly synchronised waves of contraction. You basically have very little control over how and when it contracts – it works mostly automatically.

On average, adults in their fifties who don't do much walking or other exercise lose about 150–200 grams of muscle each year.

Third, there are the voluntary muscles of your skeleton that you use to walk, run or swim, or put food into your mouth, or throw a ball.

Muscle Fibres

There are four main types of muscle fibres. They are related to the kind of work the muscle does.

The soleus muscle (the broad, flat, deep muscle of the calf) does continuous low-intensity work such as keeping you standing upright. It has lots of Type I and Type IIA fibres. They contain lots of mitochondria and can be used for a long time, but usually at a low level of intensity.

At the other extreme is the quadriceps muscle of your thigh. It's evolved for intense and rapid work – but only for a short time. It contains lots of Type IIB muscle fibres, which are low in mitochondria.

The fourth type of muscle fibre, Type IIX, is poorly understood, but shares some of the characterisations of both Type IIA and Type IIB fibres. It seems to be associated with exceptional athletic performance.

PUMPING IRON = FOUNTAIN OF YOUTH

Voluntary muscles are the ones that the body builders try to bulk up.

But the average citizen should also get some sort of resistance exercise. Overwhelmingly, the scientific and medical studies show that we should pump a little iron to keep us strong as we get older. On average, adults in their fifties who don't do much walking or other exercise lose about 150–200 grams of muscle each year. But this decline in muscle mass begins in adults even earlier – in their thirties – if they just coast along and don't do strengthening activities.

Weight Training (or Resistance Training) adds muscle bulk. If you looked at your muscles under a microscope after training, you could see the individual muscle fibres getting bigger. But Resistance Training has other effects as well.

One study showed that weight training in older people forces the muscles to remove damaged mitochondria and replace them with new ones. Mitochondria are tiny structures that make energy in virtually all of your cells. Like a furnace, they consume oxygen and make energy. And weight training in older people seems to reset the muscles back to a younger, undamaged state. It appears to be an anti-ageing treatment that rejuvenates old and damaged muscle tissue.

Muscle in Older Athletes

On average, after the age of 40 people lose some 8 per cent of their muscle mass each year. In addition, the remaining muscles get infiltrated with fat, making them weaker.

A study at the University of Pittsburgh looked at 40 competitive athletes – runners, cyclists and swimmers. There were five men and five women in each of the age groups 40–49, 50–59, 60–69 and 70+. They all were fit, training several times per week and competing frequently.

Their musculature had hardly deteriorated. The athletes in their seventies and eighties had almost as much muscle mass in their thighs as the athletes in their forties did. Furthermore, there was hardly any fat infiltration of their muscles.

Around age 60 there was a minor reduction in muscle strength, but there was hardly any further decline afterwards in their seventies.

Sports scientists once thought that as you aged, you automatically lost muscle mass. But now they know that the muscle loss is caused by inactivity, not ageing.

Studies were done with elderly rats that spent most of their lives being sedentary. These elderly rats were put on a running program. Within 13 weeks, their leg muscles had begun to develop muscle stem cells, which are known to build and repair muscle. Similar studies have not yet been done on humans.

FOOD = MUSCLE?

So you've been to the gym, you've done the weights, now what?

Consuming the right kind of nutrients immediately after a workout can change how much muscle bulk you add. People have tried all kinds of stuff, from sugar drinks to protein powders right up to illegal steroids. Steroids can indeed add muscle bulk, but have all kinds of dangerous side effects.

One surprising contender has turned out to be plain old cow's milk. One study on men showed that over a 10-week period, if they drank just two cups of skim milk after each muscle workout, they would gain almost twice as much muscle as if they drank soy drinks with the same amount of protein.

Another study looked at 20 young women. Over a 12-week period, they worked out for an hour each day, five days a week. They did leg work as well as upper-body pushing and pulling work – always under close supervision by expert trainers. After each workout, they drank a litre of either fat-free milk or a sugar (maltodextrin) drink. Compared to the sugar drinkers, the milk drinkers put on nearly double the muscle mass (an extra 1.9 kilograms of muscle over what they started with). They also lost some fat, so they didn't increase their body weight. And finally, they were stronger than the sugar drinkers.

A surprising advantage of carrying a little extra muscle is that, even at rest, it burns up more energy than fat – so muscle helps keep your weight down.

Move over, Energy Drinks – Milk Means Muscles.

Milk's Nutrition

Milk contains sugars that replace the glycogen that is lost from muscles during exercise. This sugar also stimulates the release of insulin.

Milk also contains protein. Insulin helps the protein enter the muscle, to rebuild the muscle and make it bigger.

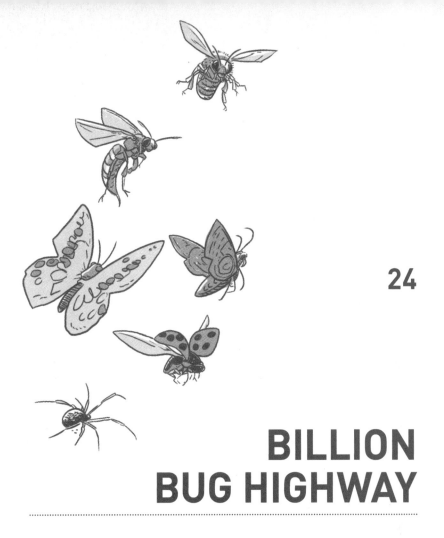

24

BILLION BUG HIGHWAY

There's an oft-repeated story that according to the Laws of Science, the bumblebee shouldn't be able to fly. The reality is that insects can fly almost as high as Mount Everest.

Butterflies can flap happily at 1 kilometre above the ground, while flies can fly to 1.5 kilometres. Midges, aphids and wasps can go a little higher, while ladybugs can reach 1.8 kilometres. Gypsy moths (mostly males looking for female partners) have been found at 3 kilometres. Spiders not using any power at all, simply floating on long, thin silk strands carried by the wind, have been seen at 4.2 kilometres. And termites have reached 6 kilometres above the ground.

You can't see them, but in the summer there are huge numbers of insects flying on many different pathways at different altitudes looking for food or a mate. Unfortunately, nature is bloody in tooth and claw – so most of them die.

How many of them are up there? Well, start by mapping out an invisible "box" that's one kilometre by one kilometre on the ground, and that reaches all the way up to the heavens. It is estimated that each month in summer some 3 billion insects would pass through this invisible box.

Bumblebees Can't Fly

There's an anti-technology sentiment that denigrates both engineers and aerodynamicists. It's the question: "Didn't an aerodynamicist prove that bumblebees can't fly?" So while it doesn't reek of personal invective or biting wit, it still hurts and it's not the whole story.

According to John McMasters, who was a former Principal Engineer on the aerodynamics staff at Boeing Commercial Aeroplanes, it seems the aerodynamicist was probably an unnamed Swiss professor famous in the 1930s and 1940s for his work in supersonic gas dynamics. The aerodynamicist was having dinner with a biologist, who idly noted that bees and wasps had very flimsy wings and heavy bodies – so how could they possibly fly?

With absolutely no hard data, but a willingness to help, the aerodynamicist did some back-of-envelope calculations, based on two assumptions.

The first assumption was that the wings were flat plates that were more or less smooth, a bit like aeroplane wings.
The second assumption was that as air flows over an insect's wings, it separates easily from the wing surface.
However, both of these assumptions were
totally incorrect.

As a result, his initial rough calculations "proved" that insects could not fly.

Of course, being a good scientist, his sense of curiosity got him interested in this problem. Clearly, he knew that insects did fly. He examined insect wings under a microscope and found that they had a ragged and rough surface, with hard hollow tubes joined by thin membranes – in other words, one of his assumptions was way off.

But by then overzealous journalists had spread the myth.
The story had flown free, even though the
bumblebee supposedly couldn't.

Bumblebees Can Fly

The aerodynamics used by flies give them the amazing
ability to do a right-angle turn in less than 50 milliseconds.
That's one-twentieth of a second, which helps explain why
they are so hard to catch.

The mystery of how insects fly was solved only
in the last 15 years.

As insects flap and rotate their wings, vortices (or spirals)
are created in the air on the leading edge of their wings. These
are called Leading-Edge Vortices or LEVs. These vortices are
robust (unlike those created by aeroplane wings) and stay
stuck to the insect wing. In turn, they produce lift, which
keeps the insect airborne.

Some insects, such as fruit flies, have a very large angle
through which they sweep their wings (145–165 degrees),
while some species of bees have more shallow strokes
(less than 130 degrees). Bees have a very high beat
frequency – 230 per second.

It turns out that bees can increase their power output
enormously. They do this by greatly increasing their
stroke amplitude to 190 degrees, but raising their beat
frequency only slightly to 235 per second. As a result, they
can fly at altitudes greater than 9 kilometres with a normal
load, or carry enormous loads that are actually greater than
their own body weight. They do this when foraging for
nectar or pollen, carrying their prey back to the hive,
or transplanting the brood to another hive.

25

PHYSICIST FIGHTS FINE (SAY THAT THREE TIMES QUICKLY . . .)

If you get a traffic fine, in the vast majority of cases it's a Fair Cop. You took on the Law, and the Law won. But every now and then, you might be genuinely innocent.

So take hope from the story of Dr Dmitri Krioukov, a physicist from the University of California, who was given a ticket for allegedly not stopping at a stop sign – and got off, thanks to Physics and Maths.

But let me warn you, this is a Tale with a Twist.

FIGHT FOR YOUR RIGHTS

In court, Dr Krioukov did not merely speak and wave his arms a lot. No, he brought with him a scientific paper that he had written based on his specific encounter with the Law – with the rather modest title of "The Proof of Innocence".

He based his case on three separate extenuating circumstances. First, that the police officer didn't measure his speed but only how quickly he swept across the officer's field of view. Second, that for the critical moment, Dr Krioukov came to a halt at the stop sign really really quickly and then took off again really quickly. Third, that at the critical moment of stopping, the police officer's view of Dr Krioukov's car was blocked by another bigger car.

> The only factor important to the police officer is whether the car actually did or did not stop at the stop sign.

SWEEP ACROSS FIELD OF VIEW

Suppose that you are standing some 30 metres away from the road, and that you have a clear view around you of 360 degrees. At first the car in the far distance seems to be moving very slowly. It has a very low angular velocity – maybe a few degrees every minute. But as it flashes past right in front of you, it seems to be moving very quickly indeed. It has an angular velocity of maybe 120 degrees in just a few seconds. It seems that the car sped up as it came past you – but this is an illusion. In reality, the car was always travelling at a fixed speed.

In the case of the alleged offence, the police officer was standing down a side street at right angles to the direction Dr Krioukov was travelling in. So as Dr Krioukov's car got closer to the stop sign, his angular velocity was increasing. To the police officer it would have appeared that Dr Krioukov's car was travelling quickly – as compared to when it was further away from the stop sign.

In reality, the only factor important to the police officer is whether the car actually did or did not stop at the stop sign. But Dr Krioukov was presenting this argument as part of his overall defence – that the police officer might have wrongly thought he was driving quickly, and could not halt at the stop sign.

GO, STOP, GO

The second circumstance is that (according to Dr Krioukov) he did come to a halt at the stop sign – but only very briefly.

According to Dr Krioukov, there was a medical reason for this. ". . . D K was badly sick with cold on that day. In fact, he was sneezing while approaching the stop sign. As a result, he involuntary (sic) pushed the brakes very hard." By a coincidence, the sneeze happened exactly as he was about to press the brake pedal.

As a result, according to Dr Krioukov's calculations, he actually decelerated very quickly, reached zero velocity at the stop sign, and then quickly accelerated back to his previous speed – all within the rather short time of 1.07 seconds.

POLICE OFFICER COULDN'T SEE

The third circumstance involves another coincidence – that during the entire length of the 1.07 seconds of hard braking, stopping and then accelerating, the police officer's view was blocked by another car.

Dr Krioukov writes "the author/defendant (D K) was driving a Toyota Yaris, which is one of the shortest cars available on the market . . . the exact model of the other car is unknown, but it was similar in length to a Subaru Outback", which is about a metre longer. If the other car had been travelling at about 19 kilometres per hour, and had been in the right place at the right time, it would have blocked off the police officer's view entirely for the whole of the 1.07 seconds (according to Dr Krioukov).

Dr Krioukov concluded: "the [police officer] made a mistake confusing the real spacetime trajectory of (D K's) car – which moved at approximately constant linear deceleration, came to a complete stop at the stop sign, and then started moving again with the same acceleration – for the trajectory of a hypothetical object moving at approximately constant linear speed without stopping at a stop sign. . . . As a result . . . the police officer's perception of reality did not properly reflect reality."

Based on all of his hard work in writing the paper, and for using Science and Maths so wisely, Dr Krioukov won the case. Or that's how the majority of the media around the world reported the story.

TWISTED TALE

However, there were a few details in the reporting that made me a little suspicious.

First, the court case was on 27 July 2011, while his paper was not published online until much later, on 1 April, AKA April Fools' Day!

Second, his paper was published on April Fools' Day.

Third, the amount of the fine that he avoided was variously reported as either $200 or $400.

So I dug a little deeper to double check.

Yes, on 16 July 2011 Dr Dmitri Krioukov did appear in a San Diego court, in front of Superior Court Commissioner Karen Riley. She said that while he did present his physics argument to her, "Much of it went right over my head". The real reason for his successful appeal was that, "The ruling was not based on his physics explanation. It was based on the officer's view . . . The officer wasn't close enough to the intersection to have a good view".

It's enough to make you think Justice is blind after all . . .

26

POPULATION
DECLINE

We humans have had an enormous effect on the planet. In any given period of time, we now shift more dirt than all the rivers on Earth. But surprisingly, our physical bodies don't take up a lot of room on the planet. All seven billion of us would easily fit into a cube 1 kilometre by 1 kilometre by 1 kilometre – and fill less than half of it.

THE WEIGHT OF PEOPLE

The heaviest people on Earth are the inhabitants of North America – 80.7 kilograms each. It doesn't sound too heavy, until you consider that it includes newborn babies, infants and children.

The North Americans are closely followed by the inhabitants of Kuwait (77.6 kilograms), Qatar (76.6 kilograms), Croatia (76.2 kilograms), United Arab Emirates (75.7 kilograms) and then Egypt (73.9 kilograms). The lightest people on Earth are the inhabitants of Bangladesh – 40.95 kilograms each.

> The heaviest people on earth are the inhabitants of North America – 80.7 kilograms each.

The average weight of a human being, spread out over the whole human race, is 62 kilograms.

The biomass of the human race (combined weight of all human flesh) is about one third of a billion tonnes. North America has just 6 per cent of the world's population, but 34 per cent of the human total biomass due to obesity. North America holds the World Record for obese people.

POPULATION GROWTH

The world population is expected to peak around 2050, when the rate of population growth should drop to zero if the current trend continues.

The population of the world would be stable if each woman had exactly two babies. Back in the 1950s, the average number was five babies, but in 2011 it had dropped to around 2.5 births per woman.

The peak population of 2050 should be somewhere between 7.5 and 10 billion people, and then it should decline. Annual births are expected to stay stable at around 134 million each year. However, deaths will increase from our current 56 million per year to about 80 million per year by 2040.

Before we "invented" agriculture, about 12,000 years ago, the world population was probably around 15 million. We reached 1 billion around 1804, 2 billion in 1927, 3 billion in 1960 – and 7 billion in 2012.

HITS TO POPULATION

But population growth is not a steady, smooth curve. Along the way, various disasters have savagely reduced the human population.

The human population is thought to have dropped down to just 2000 humans back around 70,000 BC. That was when the Toba supervolcano in what is now Indonesia erupted and cooled the world (then in an Ice Age) by about 4–5 degrees Celsius. Incredibly, all the men on Earth today almost certainly descended from just one of those men. The situation was not that this specific man said to the other 999 men, "You guys go and build a fence while I go and check out these curvy non-men creatures". No, a whole bunch of these 1000-or-so men had children who had children and so on, but all of their lines of descent died out – except for the one who fathered today's males.

According to author Matthew White, the Mongol conqueror Genghis Khan killed some 40 million people between 1206 and 1227 AD, dropping the global population by 11 per cent. The Second World War killed 66 million people, causing a 2.6 per cent population drop. In China, the An Lushan Rebellion (755–763 AD) led to the deaths of 13 million people with a 5.9 per cent drop in the global population, while the collapse of the Xin Dynasty (9–24 AD) caused 10 million deaths and a 5.9 per cent drop in the world's population.

But Mother Nature seems to hold the record for mayhem and has outdone both Genghis Khan and the Second World War. Between the years 1340 and 1400 AD, the Black Death pandemic reduced the world's population by more than 75 million from 450 million to 350–375 million (about 2.8 per cent).

27

LAZY SUN

Most of us know that the Sun does nuclear burning to generate its colossal output of energy. We also know that nuclear bombs generate a huge amount of energy from a tiny amount of mass. So how come that, weight for weight and volume for volume, the Sun puts out less energy than the compost pile in your backyard?

The Sun is a mighty beast. In 11.5 seconds, it dumps enough energy on our little planet to satisfy all our Earthly energy needs for one day. If you want to get technical, at any given moment, the Earth absorbs about 120,000 TW of power from the Sun. A TW is a trillion watts – that's a million million watts. But we humans currently generate, and use, a little more than 16 TW.

SUN 101

The Sun is a seething ball of hot gas (OK, plasma, to be strictly accurate) about 150 million kilometres away. It's about 1.4 million kilometres across.

Every second, the Sun burns 620 million tonnes of hydrogen and turns it into about 616 million tonnes of helium. That leaves about four million tonnes missing. Three million of those tonnes are turned into energy (heat, light, etc), obeying Einstein's famous equation $E=mc^2$. And one million tonnes are spewed out from that throbbing ball of hot plasma as high-energy charged particles – such as protons and electrons.

Let me give you a handle on how much mass the Sun has "burned" so far. The Sun is about 4.7 billion years old – about half-way through its life span. So far, it has lost the equivalent of about 100 times the mass of the Earth through nuclear burning. But the Sun is enormous – its mass is about 330,000 times the mass of the Earth. So 100 times the mass of the Earth hardly makes a dent.

POWER OUTPUT OF THE SUN

This nuclear burning of hydrogen doesn't happen everywhere in the Sun – no, about 99 per cent of it happens down in the central quarter (24 per cent) of the Sun. (The remaining 1 per cent happens a little further out in the next 6 per cent out from this central quarter.) The core (as it is known) is roughly 25 times bigger than the Earth. It has a temperature around 16 million degrees, and is about ten times more dense than gold or lead. Under these extreme conditions of temperature and density, when the nuclei of hydrogen atoms collide, they ultimately make a helium nucleus – and release a huge amount of energy.

The power output of the core of the Sun is about 280 watts per cubic metre – that's not quite three of the old 100 watt light bulbs.

So let's compare the power output of the Sun to both your body and a compost pile.

Depending on how you "fiddle" the statistics, you can really make the efficiency of the Sun look either "slightly bad" or "really, really bad".

OVERVIEW OF BAD/LAZY

First, to make the Sun look only slightly bad, consider the Sun's power output on a volume (cubic-metre) basis, and take only the power-emitting superdense core into account.

Second, to make the Sun look really really bad, consider the Sun's power output on a mass (kilogram) basis, and also do the sneaky trick of averaging this out over the entire volume of the Sun. It's sneaky because the overwhelming majority of the Sun does not generate power, but merely helps carry the power from the core toward the surface.

POWER PER VOLUME (SLIGHTLY BAD)

First, consider the Sun on a power per volume basis, and look only at the power-generating superdense core.

Your body emits about 100 watts of heat, and it has a volume of about one-twelfth of a cubic metre. So one cubic metre of human flesh generates about 1200 watts – about four times as much as the Sun (about 280 watts per cubic metre).

What about your backyard compost pile that takes about 180 hours to warm up from cold, reach a peak heat output, and then cool down again? That cubic metre of compost can generate an average of 6600 watts – about 23 times more than the Sun. However, compost piles have been measured generating a maximum of 14,000 watts for each cubic metre!

So when it's running flat out, volume for volume, a compost pile generates 50 times as much power as the Sun (comparing compost to the hot superdense core).

POWER PER MASS (REALLY, REALLY BAD)

Next, consider the Sun on a power per mass basis, and average the Sun's output over its entire enormous mass (not just the central core that generates the power).

One kilogram of Sun will generate a microscopic one-five-thousandth of a watt.

Your body generates more than 1 watt of power for each kilogram of mass. So your body (kilogram for kilogram) generates over 5000 times the power of the Sun.

A kilogram of veggies and grass clippings can generate a maximum of 14 watts, as it turns into organic humus in your compost pile. So (kilogram for kilogram) your compost pile can be 70,000 times more efficient than the Sun.

Mind you, your compost pile has stopped generating heat after about a week. A human body is good for about 80 years if you eat well and do exercise, and maybe even 100 years if you eat well and do exercise *and* wisely choose long-lived parents who lived to 100 years. However, our Sun will keep burning for much much longer – about 10 billion years.

WHY SO LOW?

A handful of high-grade nuclear material can destroy an entire city – once it's inside a bomb. A relatively small hydrogen bomb, such as the American W88, has the explosive power of just under half a million tonnes of TNT – and fits into a volume less than a small wheelie bin. That's 30–35 times more powerful than the nuclear bomb dropped on Hiroshima. So, if the Sun does nuclear burning, how come it generates so little power? The answer is timing and spacing.

The Sun does do nuclear burning of hydrogen nuclei, but it does it only very occasionally and sporadically. On average, any given hydrogen nucleus will fuse with another hydrogen nucleus only once every 5 billion years. Mind you, there is a huge amount of hydrogen in the Sun, so hydrogen nuclei run into each other (making nuclear reactions happen) much more frequently than once every 5 billion years.

This is why. Only a few hundred of these hydrogen-to-helium reactions happen in each cubic millimetre each second. Luckily, the core of the Sun is enormous. So even though it is only about 2 per cent as powerful as a compost pile, there's enough energy given off to keep our planet warm.

28

RADIOACTIVE CIGARETTES: X-RAY INHALE

When I was a medical doctor in the hospital system, every now and then I needed to order chest X-rays for patients who were also cigarette smokers. Quite often, they would (correctly) get nervous about the potential radiation damage from a chest X-ray. Yet when I told them that two packets of cigarettes gave them about the same radiation dose as a chest X-ray, they just would not believe me.

This raises two questions.

First, how did cigarettes get radioactive? And second, why didn't they believe me?

CIGARETTES 101

Worldwide, we humans smoke about 6 trillion cigarettes each year – enough to make a chain that would easily reach from the Earth to the Sun, and back, and then do the whole trip again, just for good measure. By the year 2020, cigarettes will be killing about 10 million people each year. They already knocked off 100 million people in the 20th century, and if things don't change they could kill one billion in the 21st century.

> By the year 2020, cigarettes will be killing about 10 million people each year.

Cigarette smoke is loaded with various chemicals that are well known to cause cancer. It's estimated that the radiation dose from the radioactive metal Polonium 210 (Po-210) in cigarettes accounts for about 2 per cent of cigarette-related deaths. That accounts for several thousand deaths each year in the USA alone.

Poisonous Polonium

Polonium-210 is extremely toxic – about 250 million times more toxic than cyanide.

Po-210 is so radioactive that it excites the surrounding air, giving off an unearthly blue glow. Weight for weight, it emits 4500 times as many alpha particles as does radium – which is notorious for being incredibly radioactive and dangerous.

A single gram of Po-210 (a cube measuring about 5 millimetres on each side) generates more heat than an old-fashioned 100 watt incandescent light bulb. That is a huge amount of power.

In 2006, Alexander Litvinenko became the first person confirmed as dying from acute Po-210 Radiation Syndrome. He had previously been an officer in the Russian KGB. After accusing his superiors of ordering the assassination of the Russian billionaire Boris Berezovsky, he escaped prosecution in Russia and was granted political asylum in the United Kingdom. In a London restaurant, he was somehow given about 10 micrograms of Po-210 – roughly 200 times more than was needed to kill him.

RADIOACTIVE CIGARETTES

Po-210 is always present whenever you have uranium. Developed countries use fertiliser that is manufactured from apatite rock, and this rock naturally contains uranium. The uranium decays to radioactive Po-210, which enters the tobacco plant through both the leaves and roots.

When a cigarette burns, it reaches temperatures of 600–800° Celsius – much hotter than the melting point of Po-210. As a result, Po-210 becomes airborne very easily. The microscopic droplets of liquefied Po-210 stick to tiny particles in the cigarette smoke. As this smoke is sucked into your lungs, the particles (carrying their toxic Po-210) then tend to land at "bifurcations" – locations in your airways and lungs where one pipe splits into two pipes.

Po-210 is intensely radioactive, and sprays alpha particles into the surrounding tissues. These alpha particles have enough energy to cause mutations and cancers.

Most people would definitely be worried if you suggested that they should have a chest X-ray every day for the rest of their lives. But some of these people quite happily smoke – sometimes up to two packets of cigarettes every day.

ALPHA PARTICLES

There are three main types of ionising radiation – alpha, beta and gamma. They can rip electrons off the atoms in your body.

Gamma radiation is electromagnetic radiation. It's like visible light, or ultra-violet – but much more powerful. It can penetrate matter easily, so it's very dangerous to life.

Beta radiation is a bunch of electrons. They are dangerous, but can be stopped by a few millimetres of metal.

Alpha radiation is a helium atom without the electrons. It's dangerous, but is even weaker than beta radiation at penetrating flesh. It has so little penetrating power that it is stopped by a few centimetres of air, or your skin. But once it gets past the dead protective layer skin, and sits on a living cell (such as in your lungs), then it can turn that lung cell into a cancer cell.

MERCHANTS OF DOUBT

It was first discovered that cigarettes contained radioactive polonium about half a century ago. So how come it's not general public knowledge?

The answer is simple: Big Tobacco is excellent at cover-ups.

These companies realised that there was radioactive Po-210 in tobacco and did their own internal, and very secret, research program. They even came up with ways to drastically reduce the amount of Po-210 in cigarette smoke. But there was no money in it. One notorious tobacco company memo about radioactive polonium read, "removal of these materials would have no commercial advantage".

> Big Tobacco's internal research showed that Po-210 was definitely harmful.

They also didn't want any bad publicity. By the early 1960s, it had already been scientifically proven that smoking was the principal cause of lung cancer. However, Big Tobacco played down the scientific truth for several decades. At the Philip Morris tobacco company, a scientist wrote in a memo to his boss, regarding research on Po-210, "it has the potential of waking a sleeping giant. The subject is rumbling [. . .] and I doubt we should provide facts."

Big Tobacco's internal research showed that Po-210 was definitely harmful. But a 1982 internal Philip Morris document advised that as long as they kept their internal research secret, they could maintain that any suggestion of a link between Po-210 and lung cancers was "spurious and unsubstantiated".

Big Tobacco companies had yet another reason for not publishing their research. Their infamous motto, as observed by a tobacco company executive in 1969, was (and still is) "Doubt Is Our Product". They used any tiny variation in the research to support the misleading claim that even the experts didn't really agree that smoking was harmful.

For example, suppose that Scientist A said that the link between smoking and lung cancer was 99 per cent certain, but Scientist B said it was 98 per cent. Big Tobacco would claim that even the scientists could not agree – and, therefore, how could anyone believe there was a link between smoking and lung cancer?

Big Tobacco plays the same deadly game today. Perhaps it's time Big Tobacco X-rayed their own internal research, so they can see through their own smoke screen.

Patriotic Polish Polonium

Polonium was discovered in 1898 by Marie and Pierre Curie. Poland had been taken over by Russia, Austria and Prussia, and effectively did not exist between 1795 and 1918.

However, Marie Salomea Sklodowska-Curie (she used both names) was both Polish-born and intensely patriotic. Not only did she teach her daughters the Polish language and take them on trips back to occupied Poland, she deliberately named the element she discovered after her home country. She did this as a political gesture, to bring attention to the cause of Poland.

She was the first woman to win a Nobel Prize, and the only woman so far to win two Nobel Prizes in two separate fields of science: Physics and Chemistry.

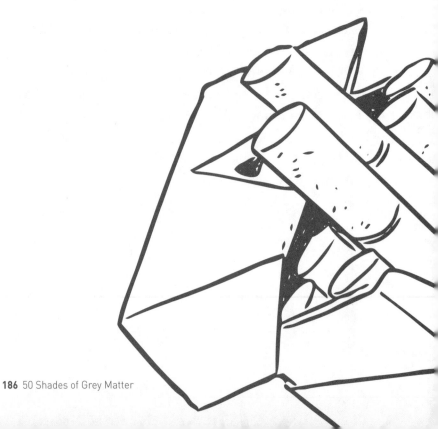

Polonium and Fertiliser

Here's a reasonable question: Polonium-210 is in
fertiliser, so it must be in our food. Why is the Po-210
in food not as dangerous as the Po-210 in tobacco?
Why doesn't food give us gut cancers?

The answer is probably that the Po-210 is "washed"
out of the gut, but not the lungs.

In cigarette smoke, it sticks to the cells at the
bifurcations in the lungs, and then radiates alpha particles.
It's always spraying radioactivity into the same cells,
increasing the chance of their turning cancerous.

Similarly, in the gut, the Po-210 in food sticks to the cells
lining the gut. However these cells have a very high turnover
rate, and are replaced every five days. Yes, a cell in the gut
might begin to turn into a cancer cell – but after five days
it will be shed, and then washed out by the continuous
flow of food, ending up in your toilet bowl.

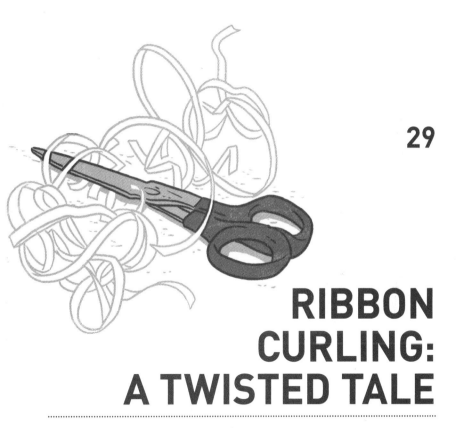

29

RIBBON CURLING: A TWISTED TALE

If you have ever bought someone a bunch of flowers, you will have almost certainly seen the florist curl the ends of the ribbon that hold the bouquet together. People have been doing this for years, but it took a physicist to understand what was going on – and to work out the best way to curl ribbons.

WHY IS IT SO?

Curling happens whenever something gets longer on one side than the other.

Dr Buddhapriya Chakrabarti from Harvard University wondered what was really going on as he saw a florist wrap up a bouquet of flowers for him. Knowing the very sensible old adage "don't reinvent the wheel", he searched both science and engineering textbooks for an explanation – and was really surprised not to find one.

Curling has been used as part of a high-speed actuation mechanism. When you heat or shine a light on these mechanisms, they curl, making part of the device move.

So, the next obvious step was to assemble a team to find out what really happens when you curl a ribbon. To do this, his team built a Motorised Curling Device. The machine had to simulate the action that you do so easily without thinking.

When you curl a ribbon, you place it on the edge of a single blade of a pair of scissors, press your thumb on it, and then pull the ribbon. The fleshy part of your thumb is soft and "flows" around the scissor blade. So the ribbon has to go around a sharp corner when you pull it.

BEST WAY TO CURL

The research team made three major findings.

First, the sharper the blade, the better the curls.

Second, the more tension that you apply to the ribbon the better – but only up to a certain point. Once you get a good curl going, you don't get any benefit from any extra tension.

The third finding was surprising. Dr Chakrabarti said, "the common intuition is that if you do it very fast you get tight curls". His team found the exact opposite.

As the ribbon is pulled between the blade and your thumb, the molecules that make up the ribbon are rearranged into a new configuration. Because they have to go around a bigger curve, the molecules on the side of the ribbon closest to your thumb are stretched more than those next to the metal blade. If you pull the ribbon through too rapidly, the molecules don't have time to settle into their new stretched-out configuration. Instead, they elastically slip back to their original configuration, and the amount of curl is less.

THE FUTURE OF CURLING

Understanding curling has surprising implications.

Human hair sometimes gets curled or straightened. But how about plants – what's going on at the micro level when the tendrils of some plants respond to being touched by curling? Doors curl (or warp) as they age – why?

With regard to human health, when malaria parasites erupt from the red blood cell they have infected, the membrane of that red blood cell curls outwards before it splits open. Understanding this specific curling might help in the treatment of malaria.

Have you heard of micromachines or nanomachines? One problem is powering them. Curling has been used as part of a high-speed actuation mechanism. When you heat or shine a light on these mechanisms, they curl, making part of the device move.

Ribbon curling is not just a complicated story – it's a twisted tale.

30

SLOW LIGHT

Most of us know that light travels really quickly. We might not know the exact speed (almost eight times around the Earth in one second), but most of us know that it's really, really fast. So how come, once light has been generated deep inside the Sun, it crawls to the surface at about one-quarter of a millimetre each second?

Let me put it another way. A photon of light takes only eight minutes to get to the Earth from the surface of the Sun. But it can take 100,000 years (OK, somewhere between 10,000 and 170,000 years) to get to the surface of the Sun after it's been generated inside the Sun's core. What's going on?

WHY SUNLIGHT IS SLOW

Well, it's a two-part answer – first, because it takes time for photons to get absorbed and re-emitted in the core, and second, this process happens many, many times.

The journey starts with hydrogen nuclei fusing together to make helium – and in the process releasing a lot of energy in the form of gamma rays. They are not yet visible light. The gamma rays can travel only a fraction of a millimetre before they're absorbed by hydrogen or helium and then re-radiated. Over and over again, they are absorbed and re-radiated. So, very slowly, the gamma rays generated by nuclear burning work their way up from the dense core.

The density of the Sun's core is incredibly high – 150 times greater than water.

The important fact is that each absorption–re-radiation event takes some time to happen – and there are lots and lots of these events.

The density of the Sun's core is incredibly high – 150 times greater than water. The atoms (actually, their nuclei – the electrons have been ripped off by the incredible temperature) are all jammed up against each other.

Simplification Warning

The whole process is way more complicated than I've outlined.

First, is the photon that leaves the surface the "same" photon that was generated in the core? Nope. Over its long travel distance, a single gamma-ray photon at very high temperature in the core gets converted into millions of visible light photons at the surface. So it's not really the "same" photon that leaves the Sun as was created 100,000 years earlier in the core. But one caused the other.

Second, "hydrogen fusing together to make helium"? Kind of. Two hydrogen nuclei fuse to make deuterium. The deuterium fuses with another hydrogen nucleus to make "light" helium, ^{3}He. And then there are four different pathways to end up at "regular" helium, ^{4}He. So at the beginning hydrogen nuclei do fuse, and at the end you do get helium – but there's a lot of messy stuff in between.

SLOWLY CLIMBING UP

After tens of thousands of years, the gamma rays climb up into what's called the Radiative Zone of the Sun's core. The Radiative Zone is huge – it stretches up from the outer core (at the 24 per cent mark) to about 70 per cent of the way to the surface. It's called the Radiative Zone because energy here travels by radiation. The gamma rays are still being absorbed and re-radiated, and each time they're being re-radiated at longer wavelengths. They gradually get converted from gamma rays down to visible light. The temperature in the Radiative Zone is around 7 million degrees Celsius near the bottom, and about 2 million degrees Celsius near the top.

Above the Radiative Zone, in the top 30 per cent of the Sun's radius, is the Convective Zone. It called the Convective Zone because energy is "convected", or carried by moving matter, just like rising water in a pot on your stove carries heat energy.

Let's look at a photon as it enters the Convective Zone, many thousands of years after it was first made at the core of the Sun. The gamma-ray photon heats the "gas" in the Convective Zone. The gas is very hot, but not quite hot enough to re-radiate the energy. Instead the hot gas rises, and it carries the energy with it. It carries energy with the hot stuff rising and coming to the surface, spreading out to the side, cooling, and then sinking again.

The photons of energy have finally, after about 100,000 years, come to the end of their journey inside the Sun. They have now reached a zone that is transparent to light. The photons escape into space and travel at the classic speed of light – around 300,000 kilometres per second. The very surface of the Sun is called the Photosphere, because that's where the photons have escaped from.

Boiling Sun

With the right telescope, you can see "boiling" patterns on the Sun's surface – the so-called "Granules".

Deeper below the surface are giant "cells" called Supergranules. They carry this hot gas towards the surface, then sideways, and then bring it down again. These Supergranules are around 30,000 kilometres across (about 2.5 times bigger than the Earth), about 10,000 kilometres deep, and last for a day or so. The hot gas moves up, across, and down relatively "slowly" – only about half a kilometre per second.

But just under the surface, the Supergranules break down into little baby Granules. Once again, in these Granules, the gas rises, spreads sideways and then sinks like water in your bubbling pot – but at around 2 kilometres per second. Granules are about 1000 kilometres in diameter, and there are over a million of them

covering the surface of the Sun. They're around
300 kilometres deep. These Granules exist for
an average of 8–20 minutes.

The Granules are hotter in the centre, and cooler at
the edges – but the average temperature at the
surface is around 5500 degrees Celsius.

SLOW START, BIG FINISH

So while in 8 minutes a photon of light can travel the 150 million
kilometres from the Sun to the Earth, in that same time a gamma
ray in the Sun's core will travel only about 10 centimetres. That's a
pretty slow rate of delivery.

The Sun is all flashy show on the surface, but deep down it's a real
homebody, slow to leave . . .

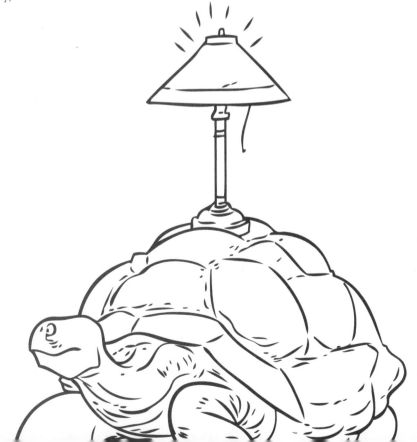

31

DIAL "D" FOR DEAREST

Men and women have close and intimate relationships that change over the decades. Around the age of 50, a man might notice that the Love of His Life doesn't ring him on the telephone as much as she used to. The guy is correct, and yes, there is someone else in her life. But that someone is not another man – it's probably their daughter.

We humans mostly try to be monogamous, and our families span several generations. So as time goes by, we change our Emotional Investment Strategies in our nearest and dearest. For example, in a woman, menopause is the end of reproduction for her. But she might have a daughter who may well reproduce.

PHONE DATA SNAPSHOT

How do you get a snapshot of people's love-lives and not get arrested as a Peeping Tom? A study by Dr Vasyl Palchykov and colleagues got their snapshot by looking at the phone records in an unnamed European country that had some 16 million mobile phone users. They accessed the phone records of one phone provider that had 6.8 million subscribers.

However, this study specifically wanted to explore relationships. So the researchers looked only at pairs of phone numbers that regularly returned each other's phone calls or text messages. This left them with 3.2 million subscribers – 1.8 million male, 1.4 million female. Over a seven-month window, these subscribers made

> Over a seven-month window, these subscribers made 1.95 billion phone calls and sent 489 million text messages.

1.95 billion phone calls and sent 489 million text messages. The data had been filtered by the phone provider so that the only information given to the researchers was the subscribers' age and gender.

The researchers analysed the data to find each subscriber's First and Second Best Friends – the people they called and texted most often.

MALE AND FEMALE BESTIES – YOUNGER

Obviously this study did have major limitations, but it did give interesting data and much food for thought.

In early adulthood, both the men and women mostly focused their attention on one member of the opposite gender (as measured by how many phone calls and texts they made). Presumably, this person was their beloved. The women began focusing their attention on one special person at a younger age than men did – 18 years for the women, and 22 for the men.

The intensity of the relationship (or at least the number of phone calls and texts) peaked at the age of 27 for the women and 32 for the men. Therefore, in this country, these are the peak ages for preferring your Best Friend to be of the opposite gender.

> The intensity of the relationship peaked at the age of 27 for the women and 32 for the men.

Interestingly, the women maintained this intensity of communication for longer than the men did – 14 years for women and 7 for the men. (Remember, this was measured by the number of phone calls and texts made by the men and women.) The authors write that "women are more focused on opposite-sex relationships than men are during the reproductively active period of their lives, suggesting they invest more heavily in creating and maintaining pairbonds than men do".

From the viewpoint of evolutionary biology, this makes sense for women wanting to have children.

DIAL "D" FOR DAUGHTER

But when women reach their forties, their First Best Friend slowly changes from being a similar-aged male to a female who is about 25 years younger.

We can't be sure without access to more of the phone company data, but this was almost certainly the adult woman's daughter. This transition of phone-calling intensity from the older male to the younger female was relatively slow and smooth. The presumed mother–daughter relationship (again, as measured by phone contacts) increased in intensity over the next 15 years, reaching a peak as the older woman reached the age of 60.

Maybe this increased communication related to the arrival of grandchildren. The new grandmother switched her time and energy from her "man" to her daughter and her grandchildren. Mother–daughter relationships are especially important in structuring human social relationships.

But what about the man who used to be that woman's First Best Friend? Well, now he is her Second Best Friend, but at least he stays that way until "death do them part". And he reciprocates by still keeping her as his First Best Friend (again, as measured by phone calls and texts).

So at around 50 years of age, both genders are finally in synchrony – because they both prefer to have a female as their New Best Friend For Ever.

32

SPACE
WEATHER

We've always had weather, but we've only had weather forecasts for a century or so. Weather forecasts can be short- or long-term. They tell us when to take a brolly, advise us when to plant crops, when to evacuate people from the path of a nasty storm, and help us plan the flight paths of planes so they can avoid headwinds and pick up tailwinds.

But now we're extending our repertoire and range by forecasting Space Weather. Long-term forecasts are now way beyond predicting whether it will rain on your birthday in a few weeks – they can now reach to the Sun. And we need these truly long-range forecasts because we're living inside the extended atmosphere and climatic influence of our active Sun.

Space Weather Affects Humans

Down here on Earth, our bodies are safe from radiation from Space. We are protected by many kilometres of atmosphere, as well as the bubble of the Earth's magnetic field.

But out in Space, people can be in Big Trouble. The Russian cosmonaut Sergei Avdeyev was terrified by very nasty Space Weather while orbiting the Earth in the space station Mir. He said, "I felt that the particles of radiation were walking through my eyes, floating through my brain, and maybe clashing with my nerves."

Quite separately from this, Mr Avdeyev has aged a little slower than normal (but read on ...). Einstein's Theory of Special Relativity tells us that time slows down as you travel faster. Mr Avdeyev has spent 747 days in space, travelling at about 27,360 kilometres per hour. So he is 20 milliseconds (that's 20 thousandths of a second, or 0.02 seconds) younger than the rest of us – if you consider only the Speed-Slows-Time effect.

However, there's the Gravity-Slows-Time effect (see "WiFi and Black Holes", page 223). Mr Avdeyev was orbiting at about 400 kilometres altitude. For you and me, about 6400 kilometres from the Earth's Centre-of-Mass, time is slowed down a bit. But for Mr Avdeyev, about 6800 kilometres from the centre of the Earth, gravity was weaker (about 90 per cent of what it is for us). So his time was a fraction faster than our time.

I leave it to the reader to work out which effect won!

WEATHER IN SPACE

That's right, we have Weather in Space. (Yup, the vacuum of Space is not absolute, so Space is not totally empty).

Our Space Weather comes from the Sun, that seething-hot star about 150 million kilometres away. Our Sun throws out about a million tonnes of charged particles every second, as well as a lot of radiation. That's the regular background level, when everything is just bubbling away quietly.

> Our Sun throws out about a million tonnes of charged particles every second, as well as a lot of radiation.

Normally, charged particles from the Sun can't make it all the way down to the surface of the Earth. They're usually deflected or slowed down by two protective layers.

First, there's the magnetic field of the Earth, which wraps around the planet like a giant bubble, about 3000 kilometres above ground level. It deflects many charged particles. Second, any remaining incoming charged particles are then also slowed by the ionospheric layer – a layer of charged particles between 85 and 600 kilometres up.

But particles that have very high energy can punch through these layers all the way down to the surface of the Earth.

High-energy particles can come from the Sun when it throws out a Coronal Mass Ejection (CME). These are bursts of millions – or billions – of tonnes of charged particles travelling at a few million kilometres per hour, at a temperature of a million degrees.

The vast majority of CMEs miss us, because the Earth is very small, and space is very big.

The number of CMEs varies with the Solar Cycle.

The Sun is not a "constant candle". It gets brighter-duller-brighter, increasing-dropping-increasing its output by 1 part per 1366, on an 11-ish year cycle. So the years 2013–14 and 2023-ish are looking to be years of maximum solar activity. In the Solar Maximum years, the Sun will have about 3–4 CMEs every day. But in the intervening years (Solar Minimums), there will be a CME only once every five days.

A century ago, when the Sun "threw a wobbly" there was no significant adverse effect upon our society. The dramatic, delightful and breathtaking aurora at the North and South Poles were pleasant windfalls.

But today it's a different story. We have satellites in orbit around the Earth, giant electric and communication grids down on the surface – and many delicate electronic toys.

SPACE WEATHER AFFECTS EARTH

Space Weather can overload the electric grid to give you a blackout.

When excited particles from the Sun hit the Earth's magnetic field, they make it fluctuate. These *magnetic* fluctuations cause *electric* current fluctuations in our electrical powerlines. These electrical fluctuations saturate the transformers in the power distribution system. Once the transformers get saturated, they overheat – and switch off. Their load is automatically transferred to other power lines and transformers, which overloads them as well. Very suddenly, the electricity can be cut off to a very large area. This happened on 13 March 1989 to the entire province of Quebec in Canada – for 9 hours. It also triggered power anomalies in the USA from the Eastern Seaboard to the Pacific Northwest.

CMEs can disrupt communication networks and make satellite navigation systems misbehave. They can fry the electronics in satellites and they can put space travellers in personal danger from the ionising radiation. Down here on the ground, they can make the flow meters in oil pipelines read wrongly. A CME can even stop your toilet flushing, by "killing" the electronics that control the water system.

Unfortunately, our systems for predicting storms in space are not very sophisticated. They are pretty crude, and there are not many of them.

Weather Forecasting on Earth

Kublai Khan, the Mongol Emperor, would have liked accurate weather forecasts for two specific years. In 1274 and 1281 he was gathering his naval forces to invade Japan. But on each occasion, the Navy of the Great Khan was destroyed by a typhoon (or at least the typhoon helped the destruction). The Japanese people believed their God had sent the kamikaze ("Kami", meaning "divine, god, spirit" and "Kaze", meaning "wind"), to protect them from Kublai Khan.

The very oldest surviving weather record covers the years 1337–1344, when William Merl of England recorded the daily weather. However he didn't have any instruments, only his natural senses such as sight, smell and hearing.

Some three centuries later, in 1663, the Royal Society of London appointed Robert Hooke to take weather records. Not only did he record the atmospheric pressure, he invented an instrument to measure wind speed. The problem was that these observations were taken at only one location in all of England, and weren't shared with anybody else.

But this all changed a few centuries later. In 1844, Samuel Morse used his electric telegraph to send messages between Baltimore and Washington, DC. In 1849, the Secretary of the Smithsonian Institute started making a simple weather chart with observations across the USA. He asked all the operators of telegraph stations to begin each morning's operations by transmitting one single word to describe the weather – such as "rain" or "stormy" or "clear".

It took another 11 years, until June 1860, before the Dutch meteorologist Christopher Buys-Ballot issued the first storm predictions in the Netherlands. He used the pressure measurements from several towns to work out the direction and speed of storms.

By 1904, the US Navy was using radio to send warnings of severe weather to ships. By 1913, they were sending daily weather bulletins to their ships. On 1 January 1921 the radio station at the University of Wisconsin made the first public daily weather forecast.

PREDICTING SPACE WEATHER

One recently invented prediction method can check the Sun for hidden CMEs. It "looks" through the Sun to "see" potential threats building up on the other side that we can't see directly.

The Sun takes about 30 days to do a complete rotation on its own axis. The Sun's hidden face is always spinning into view, and every now and then we see a potential storm come into view on the edge of the Sun. One week later, that active area (a potential CME) on the surface of the Sun has rotated further around. It is no longer on the edge of the Sun – it is now aimed straight at us, like a loaded gun. If that active area goes off and spits out a dangerous blob of radiation and charged particles at us, we are in big trouble. While our human bodies will be OK, our advanced technology won't be.

It would be extremely useful to have advance warning of major fireworks happening on the other side of the Sun before they rotate into view.

Recently, scientists have worked out a prediction method to "see" the other side of the Sun – it's called "helioseismology". Despite "helio" (from the Greek "Helios" meaning "Sun") and "seismology" being part of the name, it has nothing to do with "earthquakes on the Sun", or "Sunquakes". It just looks at up-and-down oscillations on the surface of the Sun. The turbulence inside the Sun makes it "ring" like a bell. Satellites and ground-based observatories record these oscillations.

Helioseismology works because soundwaves on our side of the Sun bring us information from the other side – and we specially want to know about stormy regions.

Now here's something really weird – sound can travel faster than light!

While light can take a hundred thousand years to get from the centre of the Sun to the surface, soundwaves can make the same trip in only a few hours. In fact, soundwaves take about seven hours to do the complete round trip from one side of the Sun to the other, and back again. They don't go through the centre of the Sun; instead they bounce around the inside of the surface. And if they bounce off an active region, which could be a source of energy and particles thrown at the Earth, the soundwaves speed up. So while the average round trip takes seven hours, if the soundwaves hit an active region they'll get back about 12 seconds sooner. The Sun takes about four weeks to rotate, so this gives us a few weeks' warning of giant solar storms. With helioseismology, we can monitor the Sun both from satellites and from the ground.

> While light can take a hundred thousand years to get from the centre of the Sun to the surface, soundwaves can make the same trip in only a few hours.

Down on Earth, weather forecasters are much maligned when they get the weather wrong. But if this technology takes off, Space Weather forecasters are on a sure bet – they can always say the weather will be sunny.

Evolution and Weather, a Sad and Silly Story

Robert Fitzroy was one of the Fathers of the Weather Forecast.

Fitzroy was the captain of the famous ship HMS *Beagle*, and travelled with Charles Darwin between 1831 and 1836. He was rewarded for his work by being made Governor of New Zealand from 1843 to 1845. However, he made himself very unpopular by defending the land rights of the New Zealand Maoris. So the English settlers quickly got him out of the way by having him recalled to England. In 1859, the now-Admiral Fitzroy was given the job of setting up a service to warn of nasty weather. He began by collecting observations from England and Ireland, which were then relayed to Lloyd's of London, where ships' captains would look at them.

In August 1861, he began making, and sending, a "forecast" to the newspapers. While the sailors and farmers loved his forecasts, some academics thought that they lacked a scientific basis.

Back when he was captain of the *Beagle*, Fitzroy had helped Charles Darwin take the observations that led to the Theory of Evolution. But Fitzroy was a religious fundamentalist, and since their time together had often spoken out against the work of Charles Darwin in the field of Evolution. However, Darwin had defenders who attacked Fitzroy in return, and they demolished him. In April 1865, Admiral Fitzroy committed suicide by cutting his throat with a razor. The Royal Society used this opportunity to stop all weather forecasts for a few years.

33

SPINACH AND POPEYE

Everyone up on western Popular Culture knows the cartoon character Popeye the Sailor Man. Popeye smokes a pipe, speaks in very mangled English and is an all-round Good Guy. He was also America's first comic book superhero (minus the shiny lycra costume). In dire straits, when mere mortals would call for Superman, Popeye turns to spinach (that glorious vegetable apparently rich in iron). For Popeye, eating a can of spinach instantly results in enormous forearms and newfound strength.

In the Land of Cartoons, everything is so simple. Spinach is rich in iron, and spinach makes Popeye strong.

Iron 101

Life loves iron, because it can both accept and donate electrons – it fits into many biochemical reactions in the body. But this very ability makes free iron dangerous in our cells, as it could lead to the production of harmful free radicals. So iron is usually bound to proteins to make it less dangerous. One such protein, ferritin, is a hollow ball, and can hold some 4500 iron atoms safely hidden inside.

Iron makes up 5 per cent of the Earth's crust, and is the fourth most common element after oxygen, silicon and aluminium.

But it's far less common in the human body, making up about 0.005 per cent (about 4.5 grams). About two-thirds is contained in haemoglobin, where it is involved in carrying oxygen from the lungs to where it's needed throughout the body. Another 5 per cent is carried inside different enzymes that control various chemical reactions, and 5 per cent is found in myoglobin in muscle for carrying oxygen. The rest is stored (in the spleen, liver and bone marrow) awaiting its incorporation into haemoglobin and other proteins.

Each day, adults need to consume about 1 milligram of iron if we are male, or 1.5–2.0 milligrams if we are female with menstrual periods. We lose our iron from sweat and by shedding cells from the skin and gut. We get our 10–20 milligrams of iron each day by eating foods such as meat, eggs, carrots, wheat, fruit, nuts, seeds and (you guessed it) green vegetables. We need to eat so much more than we need, because only 3–35 per cent of the iron we eat makes its way from the gut into our iron stores. It is mostly absorbed in the duodenum, the first part of the small intestine.

POPEYE WAS WRONG?!

But then a story spread that spinach was not rich in iron. (Ironically, this story was not true!) It wrongly claimed that the original scientists (or their printers) made a mistake, and that an exaggerated estimation of the iron content of spinach had been published.

This Urban Myth originated from an article in the *British Medical Journal* with the enticing title "Fake".

Dr T.J. Hamblin, an immuno-haematologist, wrote a piece about different fraudulent activities in

> Dr Hamblin wrote that spinach was so low in iron that Popeye would have been better off eating the can.

science and medicine. He devoted just a single paragraph out of his four-page article to the Mysterious Love Triangle of Popeye, Spinach and Iron. Dr Hamblin wrote that spinach was so low in iron that Popeye would have been better off eating the can if he wanted some iron.

Dr Hamblin stated that in the 1890s some unnamed scientists had analysed the iron content of spinach correctly. Unfortunately (according to Dr Hamblin), when the scientists wrote the result in their paper, they put the decimal point in the wrong place. They accidentally published a value for the iron content of spinach that was ten times too large. Dr Hamblin claimed that this result was eventually corrected in the 1930s by other unnamed scientists. However, all the false publicity about the iron content of spinach still continued and increased its consumption by 33 per cent during the "meatless days of the Second World War".

Dr Hamblin's main point was that unnamed 19th century scientists had led us all to the false belief that spinach was very high in iron.

But was it really false?

Unfortunately, Dr Hamblin had himself made a mistake. I unwittingly helped spread this mistake to a wider audience by reporting it.

POPEYE WAS RIGHT-ISH, AND I WAS WRONG

You see, back in the 1980s, the *British Medical Journal* had approached Dr Hamblin and asked him to write a light-hearted story. However, the mere fact of its inclusion in this well-regarded journal gave the article more credibility than it deserved.

I found this out only in 2010, when I read a paper by Dr Mike Sutton in the *Internet Journal of Criminology*. His paper was entitled "Spinach, Iron and Popeye: Ironic lessons from biochemistry and history on the importance of healthy eating, healthy skepticism and adequate citation". It was not a trivial job for Dr Sutton, a Reader in Criminology at Nottingham Trent University in the UK, to check the accuracy of Dr Hamblin's paper. Indeed, Dr Sutton spent many, many weeks doggedly searching through academic and nutrition journals dating back to the 1920s. He also read the complete set of the comic strip cartoons featuring Popeye between 1928 and 1935. Only then did he put it all together in his head, and write his 13,000-word analysis.

BAD SCIENCE X 4

Dr Sutton unearthed four main inaccuracies in Dr Hamblin's paper.

First, 19th century scientists got it wrong? No, by the end of the 19th century, they had got it right. Dr Sutton found that there was indeed a Swiss scientist who correctly analysed plants for their nutritional content.

In 1872, the German scientist Emil Theodor von Wolff had measured the iron content of spinach, and overestimated it by a factor of about 20. He measured 50 milligrams of iron per 100 grams of fresh raw spinach. (Most recently, in 2010, the United States Department of Agriculture estimated a value of 2.71 milligrams per 100 grams.) Iron is a tricky substance to measure in tiny quantities, as it exists virtually everywhere in our technological society. Trace amounts of iron can overwhelm the microscopic amounts you're trying to measure. But in 1892 (which *was* the 19th century), the Swiss scientist Gustav von Bunge got much closer to the true reading with a measurement of 4.3 milligrams per 100 grams.

Second, Popeye ate spinach for iron? Nope. The first time that he ate spinach because of its nutritional content was in a cartoon on 3 July 1932. He explains, "Spinach is full of Vitamin 'A' an' tha's what makes hoomans strong an' helty." Note that there is absolutely no mention of iron.

Third, was the increased consumption of spinach in the USA caused by Popeye? No, it had already increased massively between 1915 and 1928. This was long before Popeye began to eat it in the early 1930s.

Fourth, weight-for-weight, boiled spinach actually does contain about 50 per cent more iron than meat does – but it's mostly "non-haem iron", which is less easily absorbed. The iron in meat is mostly "haem iron", which is more easily absorbed. Haem iron makes up about 10 per cent of the iron you eat, but about 30 per cent of the iron you absorb. That means that on average the remaining 70 per cent of the iron you absorb comes from legumes, eggs, fortified cereals, and plants such as spinach and from the non-haem iron in red meat.

I GOT FOOLED

The story that spinach was wrongly reported to be rich in iron was published all over the place.

Popeye would claim that he's strong to the finish, 'cos he eats his spinach. But it had nothing to do with iron. The cartoonist behind Popeye, Elzie Crisler Segar, chose spinach for its Vitamin A content.

It took me 30 years to get my facts straight.

I'll be eating my spinach from now on with a slice of humble pie.

Spinach 101

Spinach was probably first cultivated in ancient Persia (modern Iran). It was then traded across to India and ancient China. A Chinese document claims that it came to China via Nepal, around 647 AD.

It was carried into Sicily around 827 AD, into Spain by the late 1100s, and into Germany by the 1200s. One reason for its popularity was that it grew well early in spring, before most other vegetables, offering a delicious respite from the monotony of winter vegetables.

By the 1300s, spinach was being grown France and England. It is described in the very first English cookbook, *The Forme of Cury*, in 1390.

34

WiFi AND BLACK HOLES

Generally speaking, you can think of research as fitting into one of two camps – Pure and Applied. It's fairly easy to understand Applied Research – this research is aimed at a specific goal, such as making a battery that is cheaper, lasts longer, carries more power, has fewer nasty chemicals and so on.

But what has Pure Research ever done for us?

Actually, lots of stuff, including accidentally developing WiFi while looking for black holes.

PURE RESEARCH 1: GPS

> As you go faster, mass increases, distance shrinks and time slows down. But as you move further away from a heavy mass (such as the Earth), the opposite effect happens.

Back in 1905, Albert Einstein came up with his Special Theory of Relativity. It was so abstruse that it was jokingly claimed that only three people on the whole planet could understand it.

Among other phenomena, it dealt with the effects of moving at speed and the effects of gravity. With regard to speed, as you go faster, mass increases, distance shrinks and time slows down. But with regard to gravity, as you move further away from a heavy mass (such as the Earth), the opposite effect happens.

Amazingly, these two effects described by Einstein's Theory of Special Relativity have to be accounted for to give us an accurate GPS (the Global Positioning System). If his Theory did not "adjust" for the distorted time that the GPS satellites experience (moving at 14,000 kilometres per hour and 20,000 kilometres altitude), your estimated GPS position would be wrong, and the error would increase by about 10 kilometres on the first day, 20 kilometres on the second day, and so on.

But back in 1905, and for the next half-century at least, nobody imagined that Einstein's Special Theory of Relativity could be used by slightly inebriated people on a Saturday night to help them navigate their way to a Pizza Shop to get sustenance in the form of saturated fats.

Relativity and GPS

Because GPS satellites are moving at
14,000 kilometres per hour, their "time" slows
down by about 7 microseconds every day.

Because these satellites are at an altitude of 20,000
kilometres, their "time" speeds up by about 45 microseconds
every day. (Yes, the gravitational field of our planet "slows"
down time for us Earthlings, as compared to living in no
gravitational field at all.)

When you combine these two relativistic effects (45−7=38)
you find that each day the clocks on the GPS satellites move
about 38 microseconds ahead of our clocks on the ground.
Do the maths, and the error in your location
up at about 10 kilometres each day.

Hooray for Relativity!

PURE RESEARCH 2: WWW

Particle Physicists deal with stuff smaller than atoms. Their Pure
Research gave us the World Wide Web.

These physicists needed to communicate with each other
and shift large amounts of data. So, on 25 December 1990, Tim
Berners-Lee, a computer scientist working at CERN (the European
Organisation for Nuclear Research, located just outside Geneva
in Switzerland), performed the first HTTP (HyperText Transfer
Protocol) communication between a computer and a server
connected to the internet.

At that moment, he invented the World Wide Web.

It became available to the general public on 6 August 1991. On 30 April 1993, CERN proclaimed that the World Wide Web would be made available to everybody, with no payments necessary.

PURE RESEARCH 3: BLACK HOLES TO . . .

So how did searching for black holes give us WiFi?

Back in 1974, Stephen Hawking theorised that under certain circumstances, black holes might "evaporate" – and simultaneously emit radio signals. Soon afterwards, John O'Sullivan tried to find these signals. (He's a physicist and engineer – a really unusual combination, but a good one.) His team was looking for small black holes in space – about the mass of Mount Everest, and smaller than an atom.

If these small black holes were evaporating, they would emit radio signals as they vanished. But because of their great distance from us, these signals would be hard to identify as they would be tiny by the time they arrived, and would be buried in a background of louder "noise". Furthermore, this tiny signal would be "smeared" (turned from a sharp spike into a rounded shape). So John O'Sullivan and his colleagues came up with a wonderful mathematical tool to detect these tiny, smeared signals.

As it turned out, they never did find these small black holes.

WiFi Cities

Jerusalem was the first fully WiFi-enabled city.
It was soon followed by Mysore in India in 2004.
In 2005, Sunnyvale in California became the first
US city to have free city-wide WiFi.

. . . WiFi

In 1992, John O'Sullivan was at CSIRO in Australia, trying to develop computer networks that communicated without wires.

But there was a big problem. The signals he wanted to detect were tiny, smeared and buried in a background of louder "noise".

Theoretically, there is a very expensive way to use a radio signal to send interweb access back and forth between a computer and a base station.

If they each had a powerful radio transmitter and a large parabolic dish aimed at each other – no problem. But if you had dozens of computers and base stations all trying to talk at the same time, with tiny gutless radio transmitters and tiny antennae only a few centimetres long – big problem.

His black hole mathematics turned out to be the key to WiFi. CSIRO took out patents in Australia in 1992, and in the US in 1996. By 2000, they had some working chips.

So in 2009 and in 2012, CSIRO received royalty payments of AU$250 million and AU$220 million – with probably another AU$600 million dollars to come before the patent ends. In 2009, John O'Sullivan received the Prime Minister's Prize for Science.

And now John O'Sullivan is working on the Square Kilometre Array – but that's another story, and I wonder what it will give us . . .

35

COFFEE SPILLS
AND THRILLS

Coffee is probably the world's most popular legal drug. And so, in our busy day, at some stage we will usually walk from "here" to "there" carrying a cup of coffee. And every now and then, the coffee will spill. Now this might be a surprise to you, but it took until 2012 before two engineers systematically explored this very familiar phenomenon.

COMPLICATED COFFEE CURVES

The engineers were H.C. Mayer and R. Krechetnikov from the University of California. The problem of spilling coffee is very complex, and involves two separate fields – Biomechanics and the Engineering of Sloshing Liquids.

Let's deal with Sloshing Liquids first. This field is, surprisingly, very important. Liquids that are sloshing out of control can sink a tanker ship, starve a car engine of fuel and make a liquid-fuelled rocket fail.

Now add in the Biomechanics of Walking, which cause the coffee cup to go through some very complicated motions. As you walk, your Centre of Mass follows a rather strange pathway. Your gait depends on many factors such as your gender, age, state of health and so on. After all, walking has been described as "a series of controlled falls". Your Centre of Mass is continually speeding up and slowing down in your direction of travel, as well as rising and falling – and oscillating from side to side, to boot. When you are walking, you typically rock from side to side at about 1.25 hertz, while you oscillate back and forth at around 2.5 hertz.

> When you are walking, you typically rock from side to side at about 1.25 hertz, while you oscillate back and forth at around 2.5 hertz.

But that's just the motion of your Centre of Mass.

Your cup of coffee is joined to your Centre of Mass via your hand and your wrist, elbow and shoulder joints. Each of these can move in a motion that is very different from what your Centre of Mass is doing. Your cup of coffee can tilt to the left or right, it can pitch down or up in your direction of travel, and it can even swivel to the left or right.

HEAVY ENGINEERING

The engineers drew diagrams of a walking person and of a cup, and then labelled all the relevant positions, velocities and accelerations. What they called "a frictionless, vorticity-free, and incompressible liquid", you and I would call "coffee". And we would know "an upright cylindrical container" as "a cup".

> What they called "a frictionless, vorticity-free, and incompressible liquid", you and I would call "coffee". And we would know "an upright cylindrical container" as "a cup".

And then they began to work out the Natural Resonant Frequency of the coffee oscillating in the cup.

But what is a "Natural Resonant Frequency"?

Suppose you half-fill a bathtub with water. Get a breadboard and gently pat it onto the surface of the water at one end of the bathtub – then remove the breadboard. You'll see a wave head to the other end of the bathtub. It will then bounce off and head back to your end. And it will continue back and forth, bouncing off the ends of your bathtub every few seconds. So, for water, your bathtub has a Natural Resonant Frequency of a few seconds.

Your coffee cup is much smaller, so it has a higher Natural Resonant Frequency. Depending on whether you have your coffee as an exquisite espresso or a cavernous cappuccino, the frequency might range between 4.3 and 2.6 hertz.

Typically, you pick up your cup of coffee while you are stationary, and then you accelerate. This acceleration generates the initial slosh of coffee. You continue to accelerate for a few more steps until you reach cruising speed. Typically, the initial slosh will continue to amplify until you get your first coffee spill around the sixth step.

If you concentrate (or focus) on the act of carrying
the coffee, you will also tend to accelerate more slowly
when you start walking. This slower acceleration leads to a
smaller initial slosh, which also leads to an increased number
of steps before the dreaded spill-onto-the-floor happens.

However, I have spoken to many professional waitstaff
who tell me that they get better results by not looking at all.
On the other hand, they are the professionals, far more
skilled than we amateur coffee-cup carriers.

NO MORE SPILLS

The engineers suggest a few solutions.

First, if you make the walls of the coffee cup flexible, they will absorb the energy of the incoming wave and dampen the initial slosh. Second, you could install a series of concentric rings inside the top of the cup (like egg rings). These would break down the large mass of a single slosh into a bunch of smaller masses, which would be much easier to control. A third solution would be to perforate or to drill holes in these rings. Not only would this make the rings lighter, but it would further dampen the sloshing.

Of course, you could always try the low-tech approach of "the targeted suppression of resonance frequencies", otherwise known as "walking carefully".

Other Solutions

There are other solutions, besides the obvious ones
of cup-with-a-lid, or lower-level-of-coffee-in-the-cup.

The mouth of the cup could be oval. This means that it would
have different Natural Resonant Frequencies along the long
and short dimensions of the mouth. So, while you were walking
you could slowly rotate the cup to reduce the chance of spillage.
You would have to concentrate, which might reduce your
natural enjoyment of walking back to your office
with your coffee.

The mouth of the cup could be triangular. As a wave hits
the angled walls, it would reflect, but not directly towards
the direction it came from. This could give a better
result, or worse . . .

Or you could make a little harness suspended from a string.
You would then place the cup in the harness, and swing
it like a pendulum at the end of the string.
But how nerdy do you want to look?

36

THE CARRINGTON EVENT AND THE END OF SOCIETY

Today, we humans have come a very long way from our pre-electronic ancestors. We're attached at the hip to the electronic toys that we use and enjoy – the GPS unit that finds a street in an unfamiliar city, the smartphone that is a camera and a dictionary, as well as giving access to the internet, and even the accurate watch you wear on your wrist. But just imagine: what if all our toys were suddenly to die?

Well, this is exactly what would happen in a SuperStorm, when the Sun decides to have a Hissy Fit. Welcome to the Carrington Event.

CARRINGTON

On 26 August 1859, the Sun threw a few billion tonnes of super-hot gas directly at the Earth. This was a Coronal Mass Ejection, or CME. The impact with the Earth's magnetic field and the upper atmosphere was so huge that over the next few days people saw auroras, not just near the Poles where they usually happen, but as close as within 25 degrees of the Equator. In addition, there were major hiccups in the Earth's magnetic field, and huge amounts of noise in the wires of the telegraph system – so much noise that it took 14 hours to send a mere 400 words, instead of a few minutes.

A few days later, an independently wealthy English astronomer, 33-year-old Richard C. Carrington, was partaking of his normal daily habit of observing the Sun in his private observatory. He was more than a wealthy dilettante (no slouch). As a feather in his cap, Carrington had previously discovered that the Sun rotated faster at its Equator (25 days) than at the Poles (35 days).

> The Sun rotates faster at its Equator (25 days) than at the poles (35 days).

On 1 September 1859, Richard Carrington saw something he had never seen before. He saw enormous sunspots on the Sun – so huge that they were easily visible without a telescope. Suddenly, at 11.18 a.m., they flared into an unexpected and white-hot fury. He didn't know it, but at that moment, another super-hurricane of super-hot and super-fast gas – a CME – had just been thrown at the Earth. About 17 hours later, travelling at 2380 kilometres per second, it hit our little planet.

THE CARRINGTON EVENT HITS THE EARTH

Suddenly, there were the most wonderful auroras.

Nobody had ever described auroras like these. They were so bright that people awoke at 1 a.m., thinking that the dawn was coming. The auroras generated so much light that they threw shadows, and people could read newsprint by their light in the middle of the night. The auroras got closer than 18 degrees to the Equator, being easily visible in Hawaii and Panama. A report from San Salvador (14 degrees north of the Equator) read, "The red light was so vivid that the roofs of the houses and the leaves of the trees appeared as if covered with blood."

> "The red light was so vivid that the roofs of the houses and the leaves of the trees appeared as if covered with blood."

Charged particles from the Sun almost instantly destroyed 5 per cent of the ozone in the atmosphere. The ozone layer took four years to recover. The magnetic storm set off huge currents in the ground, which invaded the long telegraph lines. Telegraph operators were nearly killed by the long, violent sparks erupting from their handsets, and several telegraph stations burned down. Even after the telegraph operators had disconnected the batteries from the telegraph lines, the currents set up by the aurora in the lines powered up the system and allowed messages to be transmitted.

CARRINGTON TODAY?

If something as big as the Carrington Event happened today, nearly 10 per cent of the 1000-or-so working satellites in orbit would stop working – that's an immediate US$100 billion cost right there. Banks rely on the super-accurate time signals from GPS satellites – so you wouldn't be able to access your money.

> Astronauts in orbit would not immediately die from acute radiation poisoning – but they would get a 70-year ("lifetime") radiation dose in just a few hours.

The vast majority of the electrical grids around the world are old, fragile and overloaded. In the USA alone, minor solar storms already cause breakdowns to the grid that increase the cost of electricity by US$500 million every 18 months. But something as big as the Carrington Event would kill the entire electrical grid of North America. Astronauts in orbit would not immediately die from acute radiation poisoning – but they would get a 70-year ("lifetime") radiation dose in just a few hours. And computers and similarly sensitive electronic equipment all over the planet would be destroyed by electrical spikes inside their delicate low-voltage circuits.

Something as huge as the Carrington Event is expected every 500 years or so. Storms half as powerful as the Carrington Event happen roughly every 50 years or so. The most recent was on 13 November 1960 – and it led to radio outages all over the world.

Big Geomagnetic Storms

The biggest geomagnetic storm for many decades
hit the Earth on 13 and 14 March 1989. The electrical
blackout in Quebec, Canada, left over five million people
without electrical power for nine hours. It cost Canada
CAD$2 billion, destroyed a US$12 million transformer, and
damaged two transformers in the UK so badly they had
to be sent back to the factory where they were made.

Back in May 1921, a geomagnetic storm burned
down a Swedish telephone exchange at Karlstad.
In New York, this storm knocked out the city's railway
line signalling system. The Carrington Event was
about 50 per cent more powerful than that storm.

The solar storm of Bastille Day in 2000 expanded
Earth's atmosphere so much that the International Space
Station, instead of losing 90–400 metres of altitude each day,
suddenly lost 15,000 metres. Indeed, the trend is that when
the Sun's activity increases, satellites fall out of orbit –
especially those in low orbits.

These CME events can also change the orbits of low-orbiting
satellites, putting them on a potential collision course
with the International Space Station.

In October and November 2003, a CME set off geomagnetic
storms that destroyed a Japanese satellite costing
US$640 million, caused blackouts in Sweden, and forced
airline flights to divert away from flying over the Arctic,
costing US$10,000 to $100,000 per flight. GPS signals
were unavailable for aircraft navigation for 30 hours.

WHAT CAN WE DO?

There are five main lines of defence against Solar SuperStorms.

The first is to learn from history. We should examine the record of magnetic data for the last 170 years as well as ionospheric data for the last 80 years. Most of this data is handwritten, and in many different locations. Basically, nobody can access this huge trove of information, but when we put this data on the web, we can then learn from it.

The second line is to understand the Physics. We still don't fully understand how CMEs travel through interplanetary space, how they inject energy into the Earth's magnetic field, and what the resulting geomagnetic storms do to our magnetic field in terms of composition, temperatures, velocities and so forth.

> We have already found a few Sun-like stars that rotate similarly to our Sun.

The third line is a little speculative, but will pay off in the long term, not the short term. We can study Space Weather in other solar systems, using, for example, the orbiting X-Ray Multi-Mirror XMM-Newton telescope. We have already found a few Sun-like stars that rotate similarly to our Sun.

Fourth, we can "harden" our satellites and our power grids down here on Earth. Already, some power utilities are installing electrical transformers that are more resistant to Earthly geomagnetic storms. Sweden, particularly vulnerable to geomagnetic variations because of its proximity to the North Pole, is hardening its entire grid.

And finally, we can look at the CMEs as they head directly towards us. In the USA, the Space Weather Prediction Center in Boulder, Colorado, gives daily space weather reports. But its budget is about one-thousandth of 1 per cent of the revenues generated by the industries it supports.

At the moment, we have only a few satellites that can give us advance warning.

The Advance Composition Explorer gives us 10–60 minutes warning of geomagnetic storms, with better than 50 per cent reliability. It floats at just 1 per cent of the distance between us and the Sun. But it's 15 years old – incredibly ancient for a satellite – and will fail one day. The new NASA Solar Terrestrial Relations Observatory (STEREO) seems to be able to give warnings up to six hours in advance.

With advance warning, we can reduce the load in various electrical power distribution grids, shut down satellites and advise astronauts to head into a shielded part of the spacecraft.

Future satellites that we send up may need to Slip, Slop and Slap to protect themselves from Solar SuperStorms . . .

37

WHAT HAPPENS TO THE RUBBER DUST FROM TYRES?

In our modern cities, roads make up about one-fifth of the urban land area, and about half of the impervious surfaces. We've driven our cars and trucks on these roads with inflatable rubber tyres for over a century. These tyres wear out and have to be regularly replaced.

Sometimes the rubber comes off in a dramatic cloud of smoke when the car skids on the road. Sometimes the road surface is sharp and slices fragments out of the rubber. But most commonly, in the course of normal rotation without skidding or cutting, the rubber is compressed and then expands. As it compresses and expands, tiny cracks develop and spread in the tread – and tiny particles of rubber flake off. Of course, this process is accelerated by the tyre being hot. The tyre can wear five times faster at 40 degrees Celsius than at 5 degrees Celsius, as the solid rubber turns into a thin liquid-like film.

Where does all this rubber go? And is it harmful?

HOW MUCH RUBBER?

There are many different ways to estimate rubber wear from a tyre.

First, each time a tyre rotates, it loses a layer of rubber about a billionth of a metre thick. (This is worked out by assuming that a tyre does about 5 million rotations before it wears out, and has lost about 5 millimetres of tread by the time it has to be replaced.) If you do some more numbers, this works out to about 4 million million million carbon atoms lost with each rotation.

Second, if physicists mount a single tyre on a test bed in a laboratory, they'll typically measure a loss of about 100 milligrams of rubber for every kilometre travelled.

Third, you can do a rough Back of Envelope calculation to get roughly the same answer.

> A busy road with 25,000 vehicles travelling on it each day will generate around 9 kilograms of tyre dust per kilometre per day.

Your average tyre weighs around 11.2 kilograms, but by the time it's balding and worn out the weight has dropped down to 9 kilograms. So each tyre loses about 2.2 kilograms, or about one-fifth of its weight, during its life. Assume that the average tyre lasts around 25,000 kilometres. Then your average vehicle with four tyres loses around 8.8 kilograms over that 25,000 kilometres. This works out to 88 milligrams of rubber lost per kilometre for each tyre – which is close to the 100 milligrams per kilometre measured in a laboratory.

If you have a busy road with 25,000 vehicles travelling on it each day, it will generate around 9 kilograms of tyre dust per kilometre per day. In America, about 600,000 tonnes of tyre dust comes off vehicles every year.

RUBBER DUST: CHEMICALS – HOW FAR AND WHAT?

Now we do know that in the Australian Outback, traces of lead from car exhausts have been found up to 50 kilometres away from the nearest road. So it's reasonable to assume that some of the tyre dust can end up a long way away – but of course, most of it will settle around the road.

Some of the tyre dust settles on and gets mashed into the road. Most of it gets blown off away from the road by the air turbulence of the vehicles. And rain easily washes the rubber dust off the road into the nearest waterways – creeks, ponds and waterlands – where it ends up as sediment on the bottom.

Tyre dust contains two main classes of chemicals – organic and inorganic.

The organic chemicals are especially toxic to aquatic creatures (such as fish and frogs) and, depending on the levels, can cause mutations or even death. They also affected human cell lines grown in glass, in a laboratory, causing an increased damage to its DNA. Latex, a component of rubber dust, has been implicated in allergies and asthma.

Some of the inorganic chemicals in tyre dust are heavy metals, such as lead and zinc. The zinc in tyres appears as zinc oxide, which is used as an activator in vulcanising sulphur to get the right structural integrity for the tyre. In fact, the presence of zinc, sulphur and silicon is a typical fingerprint for tyre dust.

RUBBER DUST: PARTICLES

There's another dark side to rubber dust – particles.

The organic and inorganic chemicals are carried as, or on, particles. In general, the smaller the particles, the more deeply they can penetrate into your lungs. PM_{10} stands for Particulate Matter that is smaller than 10 microns in size. (A micron is a millionth of a metre. A human hair is about 70 microns thick.) $PM_{2.5}$ are smaller than 2.5 microns, and are even more dangerous.

On average, about 80 per cent of all PM_{10} in cities comes from road transport. Tyre and brake wear causes about 3–7 per cent of this component. (Brake dust is rich in copper and barium.) Each year in the UK, PM_{10}s of all types are blamed for 10,000 extra deaths due to heart and lung disease.

In Europe each year, the normal wearing of tyres releases some 40,000 tonnes of Polycyclic Aromatic Hydrocarbons (PAHS), mostly as PM_{10}. PAHs are a component of the heavy oils used to make tyres. They accumulate in living tissue, and have been implicated in various cancers.

California is notorious for its heavy smog pollution – which can vary from day to day. One study ran over three years and looked at 8.7 million people, or about one-quarter of the state's population. It showed very strong links between $PM_{2.5}$ particles and the daily death rate in six Californian counties. When the $PM_{2.5}$ count was high, so was the death rate.

But we don't need to be terrified of rubber dust just yet.

We do, however, need to know how dangerous it is. Even today, after over a century of using rubber tyres, we are not still not sure of the health hazards of the rubber from the tread of tyres.

Luckily, modern tyres last much longer than they used to, so there's less tyre dust ending up in the environment.

Different Strokes

"Dirt" in the air comes from many sources – regular dust blown up by the wind of passing vehicles, or particles from wearing brakes, wearing roads and wearing tyres. The ratios vary from country to country, and between location in a country.

In Switzerland, 21 per cent of the $PM1_{10}$s in the city streets came from brake wear, 38 per cent from re-suspended road dust and 41 per cent from exhaust emissions. But the PM_{10}s in the freeways were 3 per cent brake dust, 56 per cent re-suspended road dust and 41 per cent exhaust emissions. In this case, there were no contributions from tyre wear, nor from abrasion from undamaged pavements.

Long Life Truck Tyres

Tyre engineering is full of compromises. A tyre that grips well will usually wear quickly. A tyre that gives you good fuel economy doesn't like quick turns. And so on.

Michelin claims that it can make truck tyres that last for an astonishing 640,000 kilometres. The tyres on a semi-trailer are specialised for their job – the "steer" tyres have to change direction, the "trailer" tyres have to carry the load, while the "drive" tyres have to both carry heavy loads and push the truck along the road. Michelin is trying new rubber compounds, new mechanical construction and new geometry for the grooves. In the case of a 36-tonne truck, the rolling resistance of the tyres can account for up to one-third of the fuel consumption, so the new tyres are also being engineered for lower rolling resistance.

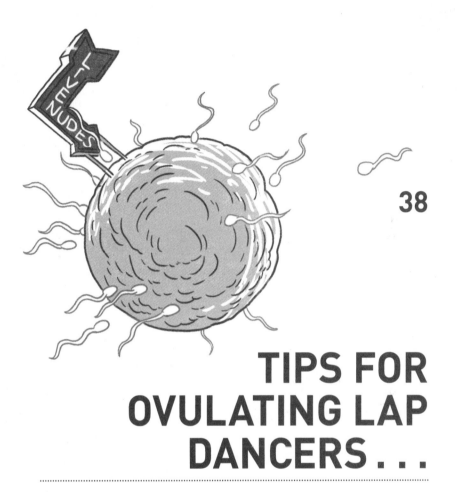

38

TIPS FOR OVULATING LAP DANCERS...

Back in July 2008, I was reading (as you do) the latest issue of the journal *Fertility and Sterility*. To my great excitement, I saw the first ever pictures of an egg actually being released over some 15 minutes from a woman's ovary. The pictures were beautiful and wonderful – and, curiously, they reminded me of an article I had read the previous year in the journal *Evolution and Human Behaviour*, about ovulating lap dancers.

Women ovulate by releasing an egg from one of their two ovaries. Clearly, ovulation is the time when a woman is most fertile and potentially most likely to become pregnant. A fertile woman has a 9.3 per cent chance of conceiving from unprotected sexual intercourse when she is ovulating. But it is not immediately obvious to most women when they ovulate. In fact, to find out exactly when they are most fertile, women can buy an Ovulation Detection Kit from a pharmacist (for $20–$60).

> A fertile woman has a 9.3 per cent chance of conceiving from unprotected sexual intercourse when she is ovulating.

Not only do most women not know exactly when they are ovulating, the vast majority of men do not consciously know when a woman is ovulating. (To be fair, it is pretty subtle. And if, true to cliché, men can't tell when a bathroom needs cleaning, or the silver needs polishing, what hope have they got?)

But, strangely, there is a small subset of "sensitive" men who somehow know when a woman is ovulating – the clients of so-called Gentlemen's Clubs.

GENTLEMEN'S CLUBS 101

Most Gentlemen's Clubs are dimly lit, serve alcohol and play music loudly. The male clients are usually between 20 and 60 years of age. They generally start the evening by getting a pile of $20 notes from the club's ATM and relaxing with a few little drinkies.

The lap dancers who work at these Gentlemen's Clubs tend to wear very little perfume, have minimal body hair, and many have breast implants. To be able to keep up with the heavy physical demands of lap dancing they do aerobic and resistance weight training, and this makes them fit and lean.

In this study of ovulating lap dancers, 18 lap dancers from Gentlemen's Clubs in Albuquerque, New Mexico, volunteered. (They were paid only $30 for volunteering.) Between them, over the two-month course of the study, they worked 296 shifts (each about 5 hours long) and did some 5300 lap dances. Seven of them were taking an oral contraceptive pill, while 11 had normal menstrual cycles. Their average age was 26.9 years, and they had been lap dancing for 6.4 years.

Gentleman?

Just a little aside. If you have never been to a Gentlemen's Club, does this imply that you're not a gentleman?

And does this further imply that the only way you can become a gentleman is by going to one of these clubs?

SHE WORKS HARD FOR THE MONEY . . .

The lap dancers did not get paid a salary by the club. Their only income was the tips. To earn their money they had to "appear ... more sexually attractive than the other five to 30 rival dancers working the same shift and [do] the 'emotional labor' of 'counterfeiting intimacy' with male club patrons".

A typical shift would run about 5.2 hours. During this time they would perform a few stage dances (perhaps every 90 minutes) to advertise themselves. For this they would receive only small tips – typically $1 to $5, and usually only from the men seated closest to the stage. This would account for only 10 per cent of their earnings.

Having "advertised" her presence, attractiveness and availability, a dancer would then spend the rest of her time walking around the club asking the men if they wanted a "lap dance". The real money would come from these three-minute lap dances, which typically cost $10 in the main club area or $20 in a more private VIP lounge. The dancer's average income from each lap dance was around $14.

To quote the paper directly, "In each lap dance, the male patron sits on a chair or couch, fully clothed, with his hands at his sides; he is typically not allowed to touch the dancer. The topless female dancer sits on the man's lap, either facing away from him (to display her buttocks, back, and hair) or facing him (either leaning back to display her breasts, and to make conversation and eye contact, or learning forward to whisper in his ear). Lap dances typically entail intense rhythmic contact between the female pelvis and the clothed male penis."

So during a typical lap dance, the male client would be close enough to the dancer for "verbal, tactile and olfactory interactions".

SHOW ME THE MONEY

On one hand, there could be no hint of a specific price for a lap dance because this would count as "illegal solicitation", violating most US state laws. But on the other hand, to make sure that the lap dancers were paid fairly, "economic norms of tips were vigorously enforced by bouncers".

Even though there was only a small sample size of dancers either on the pill or cycling normally, the results were clear. (Remember that the lap dancers who were on the pill did not ovulate, while the ones with normal menstrual cycles did ovulate.) The dancers who had normal menstrual cycles and who ovulated earned more money (an average of $276 per shift) than the ones on the pill ($193 per shift).

The results again showed a clear difference across the menstrual cycle. If the lap dancer was menstruating, she earned only $184 per shift. If she was not menstruating, but not fertile either, she earned $264 per shift. But if she was in the fertile part of the menstrual cycle (that is, ovulating), she earned the greatest sum of money – $354 per shift.

WHY? THE THEORIES

There are two main psychological schools of thought about whether ovulation is, or is not, noticeable to the opposite sex. As is often the way with psychological theories, they completely disagree with each other.

One claims that fertility or ovulation is hidden from males, while the other claims that it is revealed.

The Hidden Fertility Theory claims that it is essential for fertility to be hidden from the male. This results in the male hanging around all the time to make sure he is there when the woman is ovulating. Over time (goes this theory), his hanging-around led to bonding with the potential mother of his child, and then wanting to keep her supplied with food and shelter. Further down the line, he might become so involved with the family that he would do some "paternal care".

On the other hand, the Revealed Preference Theory claims that the male can detect when his potential partner would be most likely to fall pregnant (that is, around ovulation). In this somewhat flawed small-sample-sized (18 dancers) study of lap dancers and their clients, we do see direct economic real-world evidence in favour of this Revealed Preference Theory.

Mind you, the men's tips did not drop to zero with women who were not ovulating, so the men maintained some level of interest. Another problem with this study is that all the main measurements (tip earnings, phases of the menstrual cycle, contraception use) were self-reported by the participants over the internet to a webpage, rather then being measured by the researchers.

The voice box does respond to changing levels of female hormones. This implies that the voice is not just for speaking, but for spreading important biological information.

After all, "voice" works just fine in the dark. And it is common for us humans to have a pattern of night-time copulation.

THE VOICE

Another study that dovetails with the Revealed Preference Theory claims that a woman's voice can betray her fertility.

Dr Pipitone and colleagues from the State University of New York at Albany recorded 51 women reading out numbers (between 1 and 10) at four different times during the menstrual cycle. This included the peak fertility time at ovulation. The attractiveness of the recorded voices were rated by 34 men and 32 women. There was no real change in appeal over time for the women who were taking hormonal contra-

> There was a definite peak in attractiveness for ovulating women when they were at their most fertile, at ovulation.

ceptives. But there was a definite peak in attractiveness for ovulating women when they were at their most fertile, at ovulation. (Like some other sensitive parts of the body, the female larynx, or voice box, is supposed to be influenced by female sex hormones.)

Other studies that also support the Revealed Preference Theory claim that around ovulation women show a definite increase in verbal fluency and creativity as well as in facial attractiveness and soft tissue body symmetry. Their waist-to-hip ratio also shrinks (so a slimmer, more attractive waist combined with child-bearing hips would get the attention of a man intent upon fathering children). Ovulating women are also supposed to smell more attractive.

Other evidence suggests a human version of the Mate-Guarding Behaviour shown by the lion to his harem of lionesses. It was less intense than a bloke attacking random men with his fingernails and bare teeth if they came near his girlfriend. No, it was simply that on the days when she was at her most fertile, he would ring her more often at random times to see what she was doing! This is technically described as "higher proprietariness, attentiveness and vigilance".

With regard to lap dancers, "it seems that the optimal strategy for obtaining tips is to focus on men who are profligate, drunk, and gullible rather than those who are intelligent, handsome, and discerning". Some Prince Charming …

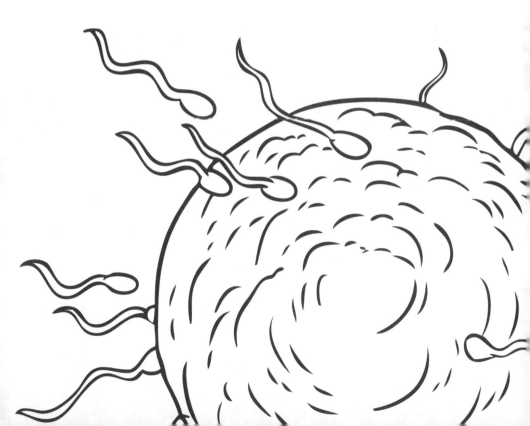

39

WHY IS THE SKY DARK AT NIGHT?

The question seems almost too obvious to ask – "Why is it dark at night?" The answers my family gave me were typically unenlightening, such as "The sun has gone to bed", and "The sun is on the other side of the Earth". Thanks, guys! But, in reality, the Question of Cosmic Darkness is very deep and subtle.

OLBERS' PARADOX

On a moonless night, the sky is dark – apart from the light of the stars. (Let's ignore the planets for now.) That starlight comes from the several hundred billion stars in our galaxy – known as the Milky Way – as well as from the stars in the several hundred billion galaxies out there in our Universe. That is a huge number of stars, and if you add them all up, they emit a lot of starlight. So why doesn't the night sky glow as brightly as the Sun?

Let's use a simple example. Imagine that you are in a clearing in a forest, and each star in the universe is a tree. You look only horizontally – not up, and not down. When you look north or south, east or west, or any direction in between, you will always see a tree trunk. In the same way, with thousands of billions of billions of stars in the sky, wherever you look, your nocturnal gaze should always land on a star – and so, the night sky should blaze with a continuous blanket of intense light. But it doesn't.

> Starlight comes from the several hundred billion stars in our galaxy – known as the Milky Way – as well as from the stars in the several hundred billion galaxies out there in our Universe.

This question has bothered astronomers for centuries, beginning with English mathematician and astronomer Thomas Digges in 1576. Many other astronomers worried about this, including Johannes Kepler, Edmond Halley, and German astronomer and physician Heinrich Olbers. The problem became known as "Olbers' Paradox".

WRONG ANSWERS

Olbers was an accomplished astronomer. In 1779 he worked out a new way to calculate the orbit of a comet, and he explained in 1811 why the tail of a comet always points away from the Sun.

With regard to Cosmic Darkness, in 1823 he suggested that most of the starlight was absorbed by gas and dust between the stars. Going back to our example of the forest, a fog would obscure the more distant trees, so that all we can see are the closest trees.

> With regard to Cosmic Darkness, Heinrich Olbers suggested that most starlight was absorbed by gas and dust between the stars.

Unfortunately, this solution is incorrect. Back in those days, scientists didn't think of light as a form of energy. They assumed that the light would be absorbed by the dust – and vanish without affecting the dust. In reality, after sufficient time, the light from more distant stars heats up the dust, and eventually brings it to the same temperature as the surface of a star, about 6000 degrees Celsius – so it too would glow.

There is another problem with Olbers' solution: the amount of dust needed would be so great that we on Earth would not be able to see our own Sun.

Other suggestions that also don't solve this problem include the theories that nearby stars block the light from distant stars, that the stars are grouped into clusters, that the Universe is expanding, and so on. (Even today, some physics textbooks wrongly claim that the darkness of the night sky is proof that the Universe is expanding. The Universe is indeed expanding, but that has nothing to do with the Dark Sky at Night problem.)

PROBABLY CORRECT ANSWERS

Edward R. Harrison gives a very nice summary of 15 possible solutions in his book, *Cosmology: The Science of The Universe*. Yes, if the Universe is infinitely old and infinitely large and has stars everywhere, then the night sky should be filled with stars. But when we look, the night sky is full of darkness, punctuated by relatively few stars.

There's a bunch of reasons why the night sky is dark.

First, the Universe that we can see is not infinitely old, but only about 13.7 billion years old. So stars have been shining for a relatively short time.

Second, the Universe is not infinitely large – the observable Universe reaches out some 46 billion light years.

Third, light takes actual time to get to us. For example, we see the Sun as it was eight minutes ago, the nearest night stars as they were four years ago, the Andromeda galaxy as it was two million years ago – and so on. This means that the light from some of the more distant stars has not yet reached us.

Fourth, stars do not shine forever. Instead, they typically burn out after several billion years. So some of the distant stars have already switched off, even though their light is still travelling towards us.

There is a fifth reason: there are not enough stars in our Universe. If there were 10 billion times more stars, this would override Reasons 1 to 4 above. But no, with our pathetic one-hundredth of a millionth of a per cent of the required number of stars, our current starlight is too feeble – by a factor of 10 billion.

THE ANSWER

So let's finish off by going back to our clearing in the forest, where we imagine each star to be a tree. We are surrounded by an inner circle of fairly old trees. Then, as we go outwards, around this inner circle are rings of progressively younger trees, and then a band of seedlings and, finally, a vast treeless plain. So the trees (and the stars) have gaps between them – which is why it's dark at night.

The problem of day and night is not as simple as black and white. It took us a couple of centuries to solve, and now we're no longer in the dark.

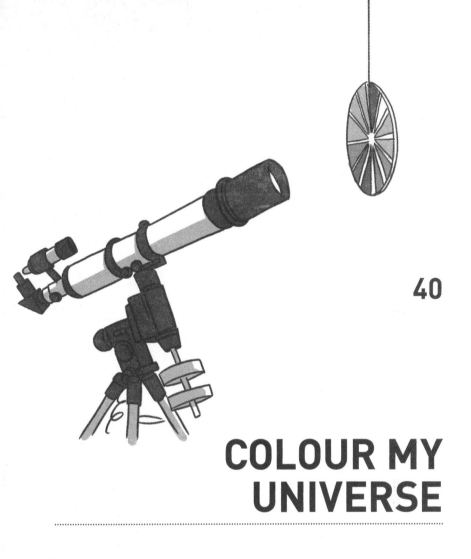

40

COLOUR MY UNIVERSE

Ever since our ancestors gathered around campfires under the stars at night, they've asked questions about the heavens. Questions such as: what are those little spots of light up there? Why are practically all of them fixed in the heavens? Why do a handful of them seem to move slowly relative to the fixed spots? How did they all get there? And where did everything come from, anyway?

Many of these questions have been answered, but one that's only recently been addressed is: What colour is the Universe? Luckily, some astronomers wondered about it and came up with an answer.

200,000 GALAXIES

It all began in Australia, when scientists strapped a fancy machine onto the top of the 3.9 metre Anglo–Australian Telescope at Coonabarabran in New South Wales. The machine is called the Two-Degree Field System (2dF) because this modern marvel will simultaneously analyse the colour of the light from 400 objects (stars and galaxies) anywhere within a 2 degree field of the sky. (This narrow sliver of sky is about 16 times bigger than the Moon as seen from Earth.) The machine has a little robot that picks up and positions optical fibres at various places across the image to detect the incoming light. It does this to an accuracy of 15 microns – or about one fifth of the thickness of a human hair.

> What colour would the Entire Universe be to a person who could stand outside it and see it all at once?

One of its first jobs was to gather the light from some 200,000 galaxies that lie within 2 billion light years of Earth. Each of these 200,000 galaxies has over 100 billion stars. These stars range in age, from very young to very old. Two scientists, Ivan Baldry and Karl Glazebrook, from Johns Hopkins University in Baltimore were part of a large team that analysed the light in many different and esoteric scientific ways.

ASTRONOMERS JUST WANNA HAVE FUN

But Karl Glazebrook decided to have some fun. What colour would the Entire Universe be to an imaginary person who could stand outside the Universe, and see all of it at once, while it was at rest?

He and Ivan Baldry assumed (quite reasonably) that these 200,000 galaxies were typical of all the 400 billion galaxies in the Entire Universe.

But how do you work out an average colour?

It's easy to measure the average height of a bunch of people. You just add up every person's height and divide the total by the number of people.

Well, it turns out that you can do the same thing with light. If you run white light into a prism, you can break up the light into many different colours. In principle, you add up all the colours, and divide by the number of colours.

Each separate pure colour has its own unique wavelength. So Glazebrook and Baldry looked at all the light coming from these 200,000 galaxies.

First, they removed the "red shift" that happens as a result of the galaxies retreating from us. (The light from a retreating galaxy looks more red than it actually is.) They then did the equivalent of adding up how much light energy there was at each particular wavelength, and divided by the number of wavelengths. It would take a huge amount of time to do these calculations and calibrations by hand, so instead they used some free software that was floating around. The average colour they got was a very pale shade of turquoise. Karl Glazebrook admitted, "It's not my favourite colour . . . on the greenish side of white."

The two astronomers wrote an amusing little footnote to their paper about how they had calculated the Colour of the Entire Universe – and used it as a story anchor in a press release. They were more than a little surprised when it grabbed the attention of the world press. Back in January 2002, the story appeared widely in all the news media around the world.

NOT TURQUOISE?

There was only one problem with this pretty picture – the astronomers were wrong.

The astronomers were very good at being astronomers – but they had no training as colour scientists.

The astronomers were very good at being astronomers – but they had no training as colour scientists. Within a few weeks, Mark Fairchild (a colour scientist from the Munsell Colour Science Laboratory at the Chester F. Carlson Center for Imaging Science at the Rochester Institute of Technology in Rochester, New York) contacted the astronomers.

Now Fairchild was a real heavy – he wrote the famous (to colour scientists) book called *Colour Appearance Models*, which discusses how colours appear to the human eye. Being a colour scientist, he quickly worked out that the astronomers' freeware computer program had automatically set a measurement called the White Point – but had set it incorrectly. The White Point is that "point" at which something looks white in your local ambient light. If you've got a sheet of white paper, it'll look white under white sunlight – but if you take it into a red-lit nightclub, it'll look red. So the White Point is different for sunlight and red light.

Fairchild worked with the two astronomers to give them the correct White Point for analysing all their colours. After they adjusted the White Point to match the perception of an observer looking at the light in a darkened room, they got a different colour of the Universe.

Suddenly, the Universe went from being a pale turquoise to beige.

Why Does the Universe have any Colour at All?

First, just to make it easy, let's ignore
the spaces between the galaxies (the black stuff).

Second, young hot stars have a blueish colour to them,
while older stars are redder. The Universe is about 13.7 billion
years old. However, the rate at which stars pop into existence
has dropped drastically over the last 6 billion years. (This is
because of a reduction in the reserves of interstellar gas,
which is needed to make new stars.) So right now, there
is an excess of older red stars. In fact, as our very pale
beige Universe grows older, it'll become redder.

Eventually all the stars will get chewed up by black holes,
and in the fullness of time the black holes will evaporate,
leaving the Universe full of nothing but light. This light
will slowly get redder over trillions and quadrillions of
years as the empty Universe continues to expand.

IF YOU DON'T MAKE A MISTAKE, YOU DON'T MAKE ANYTHING

In general, scientists are much better than most other people in our society at admitting their mistakes. Karl Glazebrook copped it on his chin, and said, "It's our fault for not taking the colour science seriously enough. I'm very embarrassed, I don't like being wrong, but once I found out I was, I knew I had to get the word out."

The astronomers released their correction to the world, causing an even bigger media feeding frenzy. They were savagely mauled for making a mistake.

> The colour beige is famous among colour theorists for having more synonyms than any other colour.

But their fellow astronomers didn't really mind.

After all, humans are not perfect and when we do make a mistake, it's good to admit it straightaway. At the time, they even set up an Astronomers' Choice Homepage for fellow astronomers to contribute names for this rather nondescript beige that apparently colours our Universe. Skyvory, Cosmic Latte and Big Bang Buff were my personal favourites.

Of course, the colour beige is famous among colour theorists for having more synonyms (50 or so) than any other colour. So you can call beige anything from almond to fallow to putty, or even wheaten.

But this leaves us with one problem. Visionaries often tell us that the only way to achieve true enlightenment is to become one with the Universe.

Does this mean we have to become one with beige . . .?

At least it will go with everything you wear . . .

Bible Meets Astronomy

The web-based *Bible Doctrine News*, which prides itself on giving a "Grace-Oriented Divine Viewpoint of Current Events", really liked the initial Universe is Turquoise report.

The editor of *Bible Doctrine News* wrote that the scientists "broke the visible light from over 200,000 galaxies in the Universe into a spectrum, or rainbow. The picture of the rainbow was emerald green with a little red from the old ones and a little blue from the young ones. This is precisely what John described when he saw the rainbow around the Throne of

God in the Control Room in Heaven."
He then quoted from Chapter 4 of *Revelation*, Verses 2 and 3:
"At once I was in the Spirit, and there before me was a throne
in heaven with someone sitting on it. And the one who sat
there had the appearance of jasper and carnelian. A rainbow,
resembling an emerald, encircled the throne."

There were two cases of misreporting here. First, the
scientists didn't actually come up with a picture of a rainbow,
just one individual colour. Second, the colour was not a rich
emerald green but a pale washed-out green. However the
editor of *Bible Doctrine News* seemed happy to "tweak" the
results to suit the message.

Then everything changed. The scientists publicly admitted
their mistake – the universe was no longer green and definitely
not a rainbow. *Bible Doctrine News* didn't mind. The editor
wrote: ". . . the new color is more reasonable, since when all
the colors of the rainbow combine, the color is white light. The
new color, although not the emerald green hue of the rainbow
around the Throne of God still agrees with the Bible. Of course,
the scientists haven't been able to see the Throne of God yet.
White is the color for sanctification. This means their mistake
has now been sanctified, or corrected."

Bible Doctrine News seems open to using any data at all
to confirm their doctrines. Maybe they will take pride in
their position on gay marriages, given their special
interests in rainbows?

41

DEATH'S WALKING SPEED

We all know the Grim Reaper, the Earthly personification of Death. This well-known mythological and literary character is easily recognisable by his extremely slim build, the black cloak with a cowl (or hoodie) that he always wears, and the long scythe that he carries.

We don't know where he lives, because electoral rolls carry only the names of the living.

But we do know how fast he walks. And if you walk faster than him, you don't die!

CHAMP

At least, if you were included in the CHAMP study. CHAMP stands for "Concord Health and Ageing in Men Project". (By coincidence, I was a medical student and junior doctor at Concord Hospital in New South Wales before this study was done.)

This study looked at 1705 men aged 70 or older. About half were born in Australia, 20 per cent in Italy, 5 per cent in Great Britain, 4 per cent in Greece and 3 per cent in China. These Sydney residents were recruited into the study over a 30-month period from January 2005 to June 2007. Since then, they have been followed for an average of about 60 months. The authors of the study documented 226 deaths by the time they published their paper in the *British Medical Journal*'s Christmas issue of 2011.

RISK STRATIFICATION OF WALKING SPEED

They found that the average walking speed of the men in their study was about 3.2 kilometres per hour, with a range of 0.5–5.8 kilometres per hour. If their preferred walking speed was less than 3 kilometres per hour, they were 1.23 times more likely to die over that 60-month period. Other studies have also shown that if you are in the slowest quarter of walkers, you are three times more likely to die in the period observed.

No walkers moving faster than five kilometres per hour died during the study.

This clearly proves that the Grim Reaper doesn't walk faster than 5 kilometres per hour. That must have been how the quicker men were able to avoid their allotted fate and outrun death. The authors write that "faster speeds are protective against mortality because fast walkers can maintain a safe distance from the Grim Reaper".

Walk Proud

It's amazing just how many diseases you will be protected from if you walk regularly – and not just the obvious ones related to leg and back strength. Walking seems to be protective against diabetes and even dementia.

In 2009, I walked across Spain with my family on the pilgrimage known as El Camino de Santiago. We walked 800 kilometres in a month and were surprised how easy it was to get into the habit of walking 8 hours per day. By comparison, anybody who has a job in which they have to stand for 8 hours per day knows how tiring just standing can be. Our evolution has made us naturally good at walking.

"Faster speeds are protective against mortality because fast walkers can maintain a safe distance from the Grim Reaper."

50 SHADES OF ACKNOWLEDGMENTS

I would like to thank all the Scientists who do real, sexy, ground-breaking Research, who refuse to submit to the Forces of Evil and spurn the Attractive Option of putting themselves into Compromising Positions – and who instead deeply seek the release gained from dealing only in Hard Facts.

In particular, I would like to thank two scientists who sternly corrected my stories for errors: Dr Greg Anderson (Spinach and Popeye) and Dr Karl Glazebrook (Colour My Universe) who willingly ran their eyes over my naked work.

Mary Dobbie and my ABC producers, Dan Driscoll and David Murray, took themselves well beyond their comfort zones to Whip these stories into their firm, taut shape. My colleague, Caroline Pegram, tied herself in knots and ravaged her Inner Goddess to finally coax out this fabulous title and Lots of other 'Secret Stuff'.

Thanks to the wonderful people from Pan. My publisher, Claire Craig, who wouldn't stop biting her lip distractingly when I fully exposed the title of the book to her. My stern editor, Emma Rafferty, who moaned deeply until I submitted … the final manuscript. The mysterious Sarah S.H. Hazelton: we never met – but you liberated my script with your hard strikes. My publicist, Jace Armstrong, who shamelessly prefers when everyone is watching. Jon MacDonald and the team from Xou Creative deliberately lay things out to drive me wild. And Douglas Holgate, whose illustrations are so alluring.

Max and Carmel Dobbie, Caroline Pegram, Sophie Hamley and my wife, Mary Dobbie, chained themselves to the Big Desk to help proof the Galleys.

Alice, you led by example and kept me wanting more … punchlines.

And Lola, you didn't need to do any more.

Thank you, all of you, for leaving your mark on this well-chiselled book. The pleasure was all mine.

REFERENCES

1. Marshmallows, Money and Munchies

"Cognitive and Attentional Mechanisms in Delay of Gratification", by Walter Mischel et al., *Journal of Personality and Social Psychology*, February 1972, pages 204–218.

"Weak Will Comes from Tired Mental Muscles", by Roy F. Baumeister, *New Scientist*, 1 February 2012.

"Don't! The Secret of Self-Control", by Jonah Lehrer, *The New Yorker*, 18 May 2009.

"Behavioural and Neural Correlates of Delay of Gratification 40 Years Later", by B.J. Casey et al., *PNAS*, 6 September 2011, pages 14,998–15,003.

2. Why is the Sky Blue?

"Human Color Vision and the Unsaturated Blue Color of the Daytime Sky", by Glenn S. Smith, *American Journal of Physics*, Vol. 73, No. 7, July 2005, pages 590–597.

"Blue-Sky Research: How Scientists Sought to Explain the Colour of the Heavens", by Richard P. Wayne, *Nature*, Vol. 330, 30 March 2009, pages 607–608.

3. Miracle Fruit

"Taste-Modifying Protein from Miracle Fruit", Kenzo Kurihara and Lloyd M. Beidler, *Science*, 20 September 1968, pages 1241–1243.

"Miraculin, The Sweetness-Inducing Protein from Miracle Fruit", J. N. Brouwer, et. al., *Nature*, 26 October 1968, pages 373–374.

"The Sweet Side of Life", Vivienne Baillie Gerritsen, *Protein Spotlight*, December 2001.

"A Tiny Fruit that Tricks the Tongue", Patrick Farrell and Kassie Bracken, *The New York Times*, 28 May 2008.

"How the Miracle Fruit Changes Sour into Sweet", by Ed Yong, *Not Rocket Science*, http://blogs.discovermagazine.com/notrocketscience/2011/09/26/ how-the-miracle-fruit-changes-sour-into-sweet, 26 September 2011.

"Human Sweet Taste Receptor Mediates Acid-Induced Sweetness of Miraculin", by Ayako Koizumi et al., *Proceedings of the National Academy of Sciences*, 28 September 2011.

4. Doorways and Forgetting

"Walking through doorways causes forgetting: Further explorations", by Gabriel A. Radvansky, Sabine A. Krawietz & Andrea K. Tamplin, *The Quarterly Journal of Experimental Psychology*, 2011, 64 (8), 24 May 2011, pages 1632–1645.

5. Breastmilk

"Maternal Defense: Breast Feeding Increases Aggression by Reducing Stress", by Jennifer Hahn-Holbrook et al. *Psychological Science*, October 2011, pages 1288–1295.

"The Wonder Of Breasts", by Florence Williams, *The Guardian*, 16 June 2012.

6. Can I Hear Seashells?

Amazing Facts About Your Body, by Gyles Daubeney Brandreth, Doubleday Books for Young Readers, New York, 1981.

"Science Q & A: Seashell Noises" by C. Claiborne Ray, *The New York Times*, 6 June 1995.

"Discovery of Sound in the Sea", http://www.dosits.org/science/soundsinthesea/commonsounds/

"She Hears Seashells: Detection of Small Resonant Cavities via Ambient Sound", by Ethan J. Chamberlain et al., *Journal of the Acoustical Society of America*, 2006, Vol. 120, No. 5, pages 3129–3129.

7. Honesty Eyes Stop Lies

"Cues of Being Watched Enhance Cooperation in a Real-World Setting", by Melissa Bateson et al., *Biology Letters*, 22 September 2006, Vol. 2, No. 3, pages 412–414.

"Feel the Eyes Upon You", by Olivia Judson, *The New York Times*, 3 August 2008.

"Dogs Know that Stealth Pays When Your Eyes are Averted", *New Scientist*, 24 July 2010, page 16.

8. Creepy App Stalks Women

" 'Girls Around Me' Dev: Our App's Not for Stalking Women, It's for Avoiding the Ugly Ones [Exclusive Interview]", by John Brownlee, 4 April 2012, http://www.cultofmac.com/158764/developers-behind-girls-around-me-stalking-app-explain-themselves-exclusive-interview/.

"Building an Online Bulwark to Fend off Identity Fraud", by Riva Richmond, *The New York Times*, 18 November 2009.

"Facebook Privacy: 10 Settings Every User Needs to Know", by Stan Schroeder, *Mashable*, 8 February 2011, http://mashable.com/2011/02/07/facebook-privacy-guide/.

"This Creepy App Isn't Just Stalking Women Without Their Knowledge, It's a Wake-Up Call About Facebook Privacy", by John Brownlee, 30 March 2012, http://www.cultofmac.com/157641/this-creepy-app-isnt-just-stalking-women-without-their-knowledge-its-a-wake-up-call-about-facebook-privacy/.

9. Death Takes a Holiday

"Holidays, Birthdays, and Postponement of Cancer Death", by Donn C. Young and Erinn M. Hade, *Journal of the American Medical Association*, 22–29 December 2012, pages 3012–3016.

10. Alcohol and Dehydration

"The Effect of Alcohol on the Renal Excretion of Water and Electrolyte", by Maurice B. Strauss et al., *Journal of Clinical Investigation*, 1955, Vol. 29, pages 1053–1057.

"Mechanism of Dehydration Following Alcohol Ingestion", by Kathleen E. Roberts, *Archives of Internal Medicine*, August 1963, Vol. 112, pages 52–55.

"Hydration Status and the Diuretic Action of a Small Dose of Alcohol", by Ruth M. Hobson and Ronald J. Maugham, *Alcohol and Alcoholism*, 24 May 2010, Vol. 45, No. 4, pages 366–373.

11. Fast For One Year

"BAD-Dependent Regulation of Fuel Metabolism and KATP Channel Activity Confers Resistance to Epileptic Seizures", by Alfredo Gimenez-Cassina et al., *Neuron*, 24 May 2012, pages 719–730.

"Features of a Successful Therapeutic Fast of 382 Days' Duration", by W.K. Stewart and Laura W. Fleming, *Postgraduate Medical Journal*, March 1973, pages 203–209.

"Fuel Metabolism in Starvation", by George F. Cahill Jr, *Annual Review of Nutrition*, 2006, Vol 26, pages 1–22.

"Running on Empty: The Pros and Cons of Fasting", by Shari Roan, *The Los Angeles Times*, 2 February 2009.

"Starving Your Way to Vigour: The Benefits of an Empty Stomach", by Steve Hendrix, *Harper's Magazine*, March 2012, pages 27–38.

12. Planet K

"The Plan to Bring an Asteroid to Earth", by Adam Mann, Wired, October 2011, http://www.wired.com/wiredscience/2011/10/asteroid-moving.

"The Tungsten Isotopic Composition of the Earth's Mantle Before the Terminal Bombardment", by Matthias Willbold et al., *Nature*, 8 September 2011, pages 195–198.

13. Cancer Creep

"The Burden of Disease and the Changing Task of Medicine", by David S. Jones et al., *The New England Journal of Medicine*, 21 June 2012, pages 2333–2338.

14. Carbon Car:
The Emission Mission

"Review of Solutions to Global Warming, Air Pollution, and Energy Security", by Mark Z. Jacobson, *Energy And Environmental Science*, December 2008.

"Preparing for a Life Cycle CO2 Measure: A Report to Inform the Debate by Identifying and Establishing the Viability of Assessing a Vehicle's Life Cycle CO_2e Footprint", by Jane Patterson et al., *LowCVP*, commissioned by Ricardo, 25 August 2011.

15. Geo-tagging Photos

"The Technology of Culture and the Military", by Paul Smith, *The Australian Financial Review*, 20 March 2012, page 30.

16. Grander Canyons

The Grand Canyon: A Different View, by Tom Vail (ed), Master Books, Green Forest, 2003.

"Seeing Creation and Evolution in Grand Canyon", by Jodi Wilgoren, *The New York Times*, 6 October 2005

"Volte-Face in the Punjab", by Philip A. Allen, *Nature*, 15 December 2005, pages 925–926.

"A Canyon-Size Age Difference: New Study Adds 11 Million Years", by Joel Achenbach, *The Washington Post*, 7 March 2008.

"Unroofing, Incision, And Uplift History Of The Southwestern Colorado Plateau From Apatite (U-Th)/He Thermochronometry", by R.M. Flowers et al., *Geological Society of America Bulletin*, May/June 2008, pages 571–587.

"The Scale of the Effect We Have on the Planet Is Yet to Sink In", by Mike Sandiford, *The Sydney Morning Herald*, May 23 2011,

"Perspective: Ingenuity Is the Test for the New Age of Man", *The Weekend Australian Financial Review*, 24 September 2011, page 54.

17. Hiccups and "Massage"

"A Phylogenetic Hypothesis for the Origin of Hiccoughs", by C. Straus et al, *BioEssays*, 2003, Vol. 25, No. 2, pages 182–188.

"Blame Tadpoles for Hiccups", by James Randerson, *New Scientist*, 8 February 2003.

"Clinical Profile and Spectrum of Commotio Cordis", by Barry J. Maron et al, *Journal of the American Medical Association*, 6 March 2002, pages 1142–1146.

"Hiccups and Digital Rectal Massage", by Majed Odeh et al., *Archives of Otolaryngology – Head And Neck Surgery*, December 1993, page 1383.

"Hiccups: A Case Presentation and Etiologic Review", by Lloyd M. Loft et al., *Archives of Otolaryngology – Head And Neck Surgery*, October 1992, pages 1115–1119.

"Ig Nobel Prizes Hail 'Digital Rectal Massage'", by Jeff Hecht, *New Scientist*, 6 October 2006.

"Linkage of Hiccup with Heartbeat", by B.Y. Chen et al., *Journal of Applied Physiology*, January 2001, pages 2159–2165.

"Termination of Intractable Hiccups with Digital Rectal Massage", by Francis M. Fesmire, *Annals of Emergency Medicine*, August 1988, page 872.

"Termination of Intractable Hiccups with Digital Rectal Massage", by Majed Odeh et al., *Journal of Internal Medicine*, January 1990, pages 145–146.

"This Old Body", by Neil H. Shubin, *Scientific American*, January 2009, pages 50–54.

"Ventricular Tachycardia as a Complication of Digital Rectal Massage", by Mark B. Lieberman, *Archives of Otolaryngology – Head And Neck Surgery*, August 1988, page 872.

18. Judges' Decisions

"Extraneous Factors in Judicial Decisions", by Shai Danziger et al., *PNAS*, 26 April 2011, pages 6889–6892.

"Judge Mental: How Bias Affects Judicial Sentences", by Jessica Hamzelou, *New Scientist*, 17 May 2012.

"No Exit: The Expanding Use of Life Sentences in America", by Ashley Nellis and Ryan S. King, *The Sentencing Project*, Washington DC, July 2009, http://www.sentencingproject.org/doc/publications/publications/inc_noexitseptember2009.pdf

"The Misuse of Life without Parole", *The New York Times*, 12 September 2011.

"The Roots of Modern Justice: Cognitive and Neural Foundations of Social Norms and their Enforcement", by Joshua W. Buckholtz et al., *Nature Neuroscience*, May 2012, pages 655–661.

19. Microwave Oven
Hertz Your WiFi

"Dueling With Microwave Ovens", by Jim Geier, http://www.wi-fiplanet.com/tutorials/article.php/3116531/Dueling-with-Microwave-Ovens.htm

20. Lunar Lunacy

"A Scale For Assessing Belief In Lunar Effects: Reliability And Concurrent Validity", by Rotton, J., Kelly, I., *Psychological Reports*, 1985, Vol. 57, pages 239–245.

"Lunar Phases and Crisis Center Telephone Calls", Wilson, J.E., Tobacyk, J.J., *Journal Society of Psychology*, 1990, Vol. 130, pages 47–51.

"Suicide and Lunar Year Review over 28 years", by S.J. Martin et al., *Psychological Reports*, 1992, Vol. 71, pages 787–795.

"The Rejection of Two Explanations of Belief in a Lunar Influence on Behavior", Angus, M.D., In: Vance, D.E. (Ed.), *Belief in lunar effects on human behavior, Psychological Reports*, 1995, Vol. 76, page 32, Unpublished masters thesis, Simon Fraser University, Burnaby, British Columbia, Canada.

"The Moon and Madness Reconsidered", by Charles L. Raison et al., *Journal of Affective Disorders*, 1999, Vol. 53, pages 99–106.

Heavenly Errors: Misconceptions about the Real Nature of the Universe, by Neil F. Comins, Columbia University Press, 2001, pages 145–149.

"Blame It on the Moon", by Gail Vines, *New Scientist*, 23 June 2001, pages 36–39.

"The Full Moon and Admission to Emergency Rooms", by Zargar M. Jhaji et al., *Indian Journal of Medical Science*, May 2004, Vol. 58 (5), pages 191–195.

"Lunacy and the Full Moon", by Scott O. Lilienfeld and Hal Arkowitz, *Scientific American Mind*, March 2009, pages 64–65.

"Biggest Full Moon", by Bob Berman, *Astronomy*, May 2012, page 12.

21. Smoking and Weightloss

"Trends in Nicotine Yield in Smoke and Its Relationship with Design Characteristics Among Popular US Cigarette Brands, 1997–2005", by Gregory N. Connolly et al., *Tobacco Control*, 2007, Vol. 16, No. e5.

"Nicotine Decreases Food Intake Through Activation of POMC Neurons", by Yann S. Mineur et al., *Science*, 10 June 2011, Vol. 332, pages 1330–1332.

22. The Anne Hathaway Effect

"In Class Warfare, Guess Which Class Is Winning", by Ben Stein, *The New York Times*, 26 November 2006.

"Stock Traders Find Speed Pays, in Milliseconds", by Charles Duhigg, *The New York Times*, 24 July 2009.

"Fast Traders, in Spotlight, Battle Rules" by Graham Bowley, *The New York Times*, 17 July 2011.

"The Hathaway Effect: How Anne Gives Warren Buffett a Rise", by Dan Mirvish, *The Huffington Post*, 2 March 2011.

"Analysing the Market Doesn't Quite Compute", by Stephen Shore, *The Australian Financial Review*, 14 May 2011, page 11.

23. Milk Magnifies Muscle

"The Transcriptional Coactivator PGC-1ß Drives the Formation of Oxidative Type IIX Fibers in Skeletal Muscle", by Zoltan Arany et al., *Cell Metabolism*, January 2007, Vol. 5, pages 35–46.

"Consumption of Fluid Skim Milk Promotes Greater Muscle Protein Accretion After Resistance Exercise than Does Consumption of an Isonitrogenous and Isoenergetic Soy-Protein Beverage", by Sarah B. Wilkinson et al., *American Journal of Clinical Nutrition*, April 2007, Vol. 85, pages 1031–1040.

"Body Composition and Strength Changes in Women with Milk and Resistance Exercise", by Andrea R. Josse et al., *Medicine & Science in Sports & Exercise*, June 2010, Vol. 42, No. 6, pages 1122–1130.

"Ageing Well Through Exercise", by Tara Parker-Pope, *The New York Times*, 9 November 2011.

24. Billion Bug Highway

"The Flight of the Bumblebee and Related Myths of Entomological Engineering: Bees Help Bridge the Gap Between Science And Engineering", by John H. McMasters, *American Scientist*, March/April 1989, pages 164–169.

"The Aerodynamics of Free-Flight Manoeuvres in Drosophila", by Steven N. Fry, *Science*, 18 April 2003, pages 495–498.

"Leading-Edge Vortex Lifts Swifts", by J.J. Videler, *Science*, 10 December 2004, pages 1960–1962.

"Short-Amplitude High-Frequency Wing Strokes Determine the Aerodynamics of Honeybee Flight", by Douglas L. Altshuler et al., *PNAS*, 13 December 2005, pages 18213–18218.

"The Invisible Highway" by Benjamin Arthur and Robert Krulwich, *NPR*, http://www.youtube.com/watch?v=-QxfOYhpjro.

25. Physicist Fights Fine

"The Mathematical Proof that Got a Physicist out of a Traffic Ticket", by Alasdair Wilkins, io9.com/5902182/the-mathematical-proof-that-got-a-physic.

"The Proof Of Innocence", by Dmitri Krioukov, arXiv:1204.0162v1 [physics.pop-ph], 1 April 2012, http://arxiv.org/pdf/1204.0162v2.pdf.

26. Population Decline

"The Weight of Nations: An Estimation of Adult Human Biomass", by Sarah C. Walpole et al., BMC Public Health, 2012, Volume 12:439.

"Population Control, Marauder Style", The New York Times, 5 November 2011.

The Great Big Book of Horrible Things: The Definitive Chronicle of History's 100 Worst Atrocities, by Matthew White, W.W. Norton and Company, New York, 2011.

27. Lazy Sun

"Optimisation of In-Vessel Co-Composting through Heat Recovery", by Maurice Viel et al., Biological Wastes, March 1987, Vol. 20, pages 167–185.

Why Does E=mc2 (And Why Should We Care), by Brian Cox and Jeff Forshaw, Da Capo Press, Boston, 2010, pages 171–218.

"How Clean is Green?", by Anil Ananthaswamy, New Scientist, 28 January 2012, pages 34–38.

28. Radioactive Cigarettes

"Polonium–210: A Volatile Radioelement in Cigarettes", by Edward P. Radford Jr and Vilma R. Hunt, Science, 17 January 1964, pages 247–249.

"Puffing on Polonium", by Robert N. Proctor, The New York Times, 1 December 2006.

"Waking a Sleeping Giant: The Tobacco Industry's Response to the Polonium-210 Issue", by Monique E. Muggli et al., American Journal of Public Health, September 2008, pages 1643–1650.

"The Polonium Brief: A Hidden History of Cancer, Radiation, and the Tobacco Industry", by Brianna Rego, Isis, September 2009, pages 453–484.

"Radioactive Smoke", by Brianna Rego, Scientific American, January 2011, pages 78–81.

29. Ribbon Curling

"Secret of Ribbon Curling Revealed: For Tightly Curled Ribbons, Pull the Scissors Slowly", by J.R. Minkel, Scientific American, 14 February 2007, http://www.scientificamerican.com/article.cfm?id=secret-of-ribbon-curling.

"Self-Similar Curling of a Naturally Curved Elastica", by A.C. Callan-Jones et al., Physical Review Letters, 27 April 2012, pages 174302-1–174302-5.

30. Slow Light

"On the Photon Diffusion Time Scale for the Sun", by R. Mitalas and K. Sills, The Astrophysical Journal, 20 December 1992, Vol. 401, pages 759–760.

"Ancient Sunlight", Technology Through Time, Issue 50, 2007, NASA http://sunearthday.nasa.gov/2007/locations/ttt_sunlight.php

31. Dial "D" for Dearest

"Sex Difference in Intimate Relationships", by Vasyl Palchykov et al., Nature Scientific Reports, 19 April 2012, Vol 2: 370, DOI: 10.1038/srep00370.

"Phone Data Shows Romance 'Driven By Women'", by Pallab Ghosh, BBC News, 19 April 2012, http://www.bbc.co.uk/news/science-environment-17729478.

"BFF?: Cell Phone Study Shows Evolving Lifetime Relationships in Men and Women", by Daisy Yuhas, 20 April 2012, http://www.scientificamerican.com/search/?q=BFF%3F%3A+Cell+Phone&x=0&y=0

32. Space Weather

"Light Fantastic", by Stuart Clark, New Scientist, 31 May 2008, pages 39–41.

"Fitzroy, Robert", Encyclopaedia Britannica, Encyclopaedia Britannica Ultimate Reference Suite 2011, Chicago, DVD.

"How Clean Is Green?", by Anil Ananthaswamy, New Scientist, 28 January 2012, pages 34–38.

33. Spinach and Popeye

"Fake", by T.J. Hamblin, British Medical Journal, 19–26 December 1981, pages 1671–1674

"Further Studies on the Availability of Iron in Biological Materials", by W.C. Sherman, C.A. Elvehjem and E.B. Hart, Journal of Biological Chemistry, Vol. 107, No. 3, 1934, pages 383–394.

"Modifications of the Bibyridine Method for Available Iron", by G. Kohler, C. Elvehjem and E. Hart, Journal of Biological Chemistry. Vol. 113, 1936, pages 49–53.

"Spinach, Iron and Popeye: Ironic lessons from biochemistry and history on the importance of healthy eating, healthy skepticism and adequate citation", Dr Mike Sutton, Internet Journal of Criminology, March 2010

34. WiFi and Black Holes

"Real-World Relativity: The GPS Navigation System", access date 12 June 2011, http://www.astronomy.ohio-state.edu/~pogge/Ast162/Unit5/gps.html

"How Australia's Top Scientist Earned Millions from Wi-Fi", by David Sygall, *The Sydney Morning Herald*, 7 December 2009.

35. Coffee Spills and Thrills

"Walking With Coffee: Why Does It Spill?", by H.C. Mayer and R. Krechetnikov, 26 April 2012, *Physical Review Letters*, Vol. 85, 046117, pages 046117-1–046117-7.

36. The Carrington Event

"The 1859 Solar–Terrestrial Disturbance And The Current Limits Of Extreme Space Weather Activity", by E.W. Cliver and L. Svalgaard, *Solar Physics*, 2004, Vol. 224, pages 407–422.

"Forecasting the Impact of an 1859-Calibre Superstorm on Satellite Resources", by Sten Odenwald, James L Green and William Taylor, *Advances in Space Research*, Vol. 38, No. 2, 2005, pages 280–297.

Committee on the Societal and Economic Impacts of Severe Space Weather Events: A Workshop, National Research Council, *Severe Space Weather Events – Understanding Societal and Economic Impacts, A Workshop Report*, The National Academies Press, Washington, 2008.

"Bracing for A Solar Superstorm", by Sten F. Odenwald and James L. Green, *Scientific American*, August 2008, pages 80–87.

"Sun Storms" by Richard A. Lovett, *Cosmos*, February 2002, pages 59–67.

"Prepare for the Coming Space Weather Storm", by Mike Hapgood, *Nature*, 19 April 2012, pages 311–313.

37. What Happens to Rubber Dust?

"Burning Rubber" by Fred Pearce, *New Scientist*, 10 April 1999.

"Impact of tire debris on *in vitro* and *in vivo* systems", by Maurizio Gualtieri et al., *Particle and Fibre Toxicology*, 24 March 2005, Vol. 2, No. 1.

"A Case Study of Tire Crumb Use on Playgrounds: Risk Analysis and Communication When Major Clinical Knowledge Gaps Exist", by Mark E. Anderson et al., *Environmental Health Perspectives*, January 2006.

"Characterization of Metals Emitted from Motor Vehicles", by James J. Schauer et al., *Research Report*, Health Effects Institute, No. 133, March 2006.

"The Effects of Components of Fine Particulate Air Pollution on Mortality in California: Results from CALFINE", by Bart Ostro et al., *Environmental Health Perspectives*, January 2007, pages 13–19.

"A Truck Tyre That Goes The Extra 100,000 Miles", by Stuart F. Brown, *The New York Times*, 26 August 2007.

"Occurrence and effects of tire wear particles in the environment – A critical review and an initial risk assessment", by Anna Wik and Göran Dave, *Environmental Pollution*, January 2009, pages 1–11.

"Theory of Powdery Rubber Wear" by B.N.J. Persson, *Journal of Physics: Condensed Matter*, 2 December 2009, Vol 21, 485001.

"PM10 emission factors for non-exhaust particles generated by road traffic in an urban street canyon and along a freeway in Switzerland", by N. Bukowiecki et al., *Atmospheric Environment*, June 2010, pages 2330–2340.

38. Tips for Ovulating Lap Dancers ...

"Acoustic Measure of Hormone Affect on Female Voice During Menstruation", by Larry Barnes et al., *International Journal of Humanities and Social Science*, March 2011, pages 5–10.

"Fertile Women Betrayed by Voice", *New Scientist*, 3 May 2008, page 14.

"Ovulatory Cycle Effects on Tip Earnings by Lap Dancers: Economic Evidence for Human Estrus?", Geoffrey Miller et al., *Evolution And Human Behaviour*, November 2007, pages 375–381.

"Women's Voice Attractiveness Varies across the Menstrual Cycle", by R. Nathan Pipitone et al., *Evolution and Human Behaviour*, July 2008, pages 268–274.

39. Why is the Sky Dark at Night?

Cosmology: The Science Of The Universe, Edward R. Harrison, Cambridge University Press, 1986, pages 249–265.

"The Extra-Galactic Background Light A Modern Version Of Olbers' Paradox", Paul S. Wesson, *Space Science Reviews*, 1986, Vol. 44, pages 169–176.

"Johann Mädler's Resolution Of Olbers'
Paradox", F.J. Tipler, *Quarterly Journal
Review Of The Astronomical Society*, 1988,
Vol. 29, pages 313–325.

"The Dark Night Sky Riddle – Olbers'
Paradox", Edward Harrison, *The Galactic
And Extragalactic Background Radiation:
Proceedings of the 139th Symposium of the
International Astronomical Union*, 1990,
pages 3–17.

Introduction To Cosmology, Matts Roos,
3rd ed, 2003, John Wiley and Sons,
pages 9–11.

40. Colour My Universe

"Green Is the Colour", by Eugenie Samuel,
New Scientist, 19 January 2002, page 16.

"The Universe Is Not Turquoise – It's Beige",
by Eugenie Samuel, *New Scientist*,
7 March 2002.

"The 2dF Galaxy Redshift Survey:
Constraints on Cosmic Star-Formation
History from the Cosmic Spectrum", by Ivan
Baldry, Karl Glazebrook et al., Astrophys. J.
569:582, 2002, arXiv:astro-ph/0110676v2.

"The Cosmic Spectrum and the Colour of
the Universe", by Karl Glazebrook and Ivan
Baldry, 28 December 2004,
http://www.pha.jhu.edu/~kgb/cosspec/

41. Death's Walking Speed

"How Fast Does the Grim Reaper Walk?
Receiver Operating Characteristics Curve
Analysis in Healthy Men Aged 70 and Over",
by Fiona F. Stanaway et al., *British Medical
Journal*, 15 December 2011.

"Gait Speed and Survival in Older Adults",
by Stephanie Studenski et al., *Journal of the
American Medical Association*, 5 January
2011, pages 50–58.

"Role of Gait Speed in the Assessment of
Older Patients", by Matteo Cesari, *Journal of
the American Medical Association*,
5 January 2011, pages 93–94.